Vegetables
you used to hate!

Vegetables
you used to hate!

Darlene King

Whitecap Books
Vancouver/Toronto

Edited by Elaine Jones
Proofread by Elizabeth McLean
Cover design by Susan Greenshields
Interior design by Margaret Lee
Photography by Christian Lacroix
Art direction by Pierre Durand
Food styling by Lise Carrière

Printed and bound in Canada.

Canadian Cataloguing in Publication Data

King, Darlene,
     Vegetables you used to hate

       Includes index.
       ISBN 1-55285-061-7

       1. Cookery (Vegetables). I. Title.
TX801.K56 2000   641.6'5   C00-910068-7

The publisher acknowledges the support of the Canada Council and the
Cultural Services Branch of the Government of British Columbia in making
this publication possible. We acknowledge the financial support of the
Government of Canada through the Book Publishing Industry Development
Program for our publishing activities.

For more information on other Whitecap Books titles,
please visit our web site at www.whitecap.ca

To my family, who instilled in me at a very young age that the dinner table was a place for coming together and sharing conversation as well as food. And to my husband, Tom, and our many friends with whom we carry on this tradition.
Santé.

ACKNOWLEDGMENTS

I would like to express a heartfelt thank you to a number of people for their support in the writing of this book.

First, to my husband, Tom: your belief in me and enthusiasm for all my endeavors is an inspiration daily.

To Lise Carrière for her artistry in food styling and Christian Lacroix for his beautiful photography: I've told you both before and I'll tell you again—you always make me look so good. It's such a pleasure working with you.

Thanks to Lynne Timmermanis, my first cooking instructor and role model, some of whose recipes appear in this book—Thai Cold Spring Rolls and Caramelized Onion, Shallot and Garlic Tart; to Diane Foster for her contribution, Rhubarb Sour Cream Pie; and lastly to Mary Lou Smith—we started on this path together and I thank you for your contributions and encouragement as well.

Maureen Grice and the International Olive Oil Council offered help and contributed the Mediterranean Spinach and Feta Stuffed Bread and Mediterranean Stuffed Peppers.

Finally, thanks to my sisters, Debbie and Donna, and to my mom, Joan, for their many phone calls of encouragement.

CONTENTS

INTRODUCTION

Vegetables have taken a back seat to the meats and sweets presented on our dinner tables for generations, but today's cooks want more on their plate. We hated vegetables as kids —that mushy mash that we were expected to swallow—but that's all behind us now. Not only are vegetables good for you, they give a plate texture, form, color and, most importantly, taste.

I've always been a foody, as long as I can remember. My earliest memories are what I had for dinner at my sixth birthday party, many years ago. Nowadays, I remember places and vacations by the restaurants I've eaten in and will recall the entire menu of a particularly memorable evening years later.

I love everything about vegetables. I love to shop for them. I enjoy squeezing the tomatoes for just the right resistance. The sheen of a meaty purple eggplant can send me right across the produce aisle. And the sight of the first asparagus spears peeking through the soil each spring—well, let's leave that one alone.

The intoxicating melange of produce on view at a farmer's market will draw me right in. Deciding what I'll prepare with a new variety of squash or a ripe white cauliflower might initiate a lively conversation with the farmer on how to best prepare a dish I've never tried before. I love to share ideas, recipes and customs of food, especially with other foodies.

I love to cook, and I'll spend countless hours poring over ideas for a dinner party. But my favorite part is cooking for my family and friends. I know no greater pleasure than watching someone who claims to hate vegetables take that first mouthful. Invariably, their faces light up with a smile that radiates surprise and pure pleasure. When asked for the secret, I reply, "These are vegetables, the way they were meant to be."

If my philosophy seems a little sensual, that's because good food calls all our senses to attention. The preparation of a meal is heartfelt, a gift from the cook to all those sharing in it. Possibly even more important to serious cooks, it is the palette of their art and an expression of love.

When it comes to ingredients, there are two or three villains—namely, butter, salt and cream. But these are also the cornerstones of good food, so don't be afraid to use them. Cut back a little in the amounts asked for, rather than substituting milk or margarine. They are the essence of fabulous food and that's all I'm interested in cooking. It's my creed.

With any luck this book will encourage you to go out and grab a few onions, mess around with a couple of hot peppers and put you in the mood for some fun with veggies.

Bon appétit!

DIPS, SPREADS AND APPETIZERS

*Appetizers were originally served as savory
little morsels presented before the main course.
Their purpose was to stimulate the appetite and
make you hungry for the rest of the meal.*

*Appetizers are growing in popularity, and
this trend is most commonly seen in the growing
selection of choices now available in restaurants.
People like to try an assortment of different dishes
served in smaller portions. This allows them to
sample differing colors, textures and global tastes.
It's now quite common to combine two or three different
appetizers in a meal, instead of having an entrée.*

Hot Baked Artichoke Dip

1	loaf pumpernickel or rye bread, unsliced	1
1	12-oz. (340-mL) jar marinated artichoke hearts, drained	1
1 cup	grated Parmesan cheese	240 mL
1/3 cup	mayonnaise	80 mL
2/3 cup	plain yogurt	160 mL
1	small clove garlic, crushed	1
1/4 tsp.	lemon juice	1.2 mL
1/4 tsp.	hot pepper sauce	1.2 mL
	salt and freshly ground black pepper to taste	
1/4 cup	finely chopped fresh parsley	60 mL

Serves 8

This is a great dish for your buffet table or a casual gathering of friends. Marinated artichoke hearts make it so easy to prepare. Have a second loaf on hand for extra dipping. This dip is also great with fresh vegetables. Try fresh endive—the pieces can be used to scoop up the dip and they provide a nice crunchy texture.

Prepare the loaf by slicing off the top. Scoop out the interior and cut it into 1-inch (2.5-cm) squares. Set the bread aside while making the filling.

Preheat the oven to 375°F (190°C).

Chop the artichoke hearts. Mix the cheese, mayonnaise, yogurt, garlic, lemon juice, hot pepper sauce, salt and pepper until well combined. Add the artichoke hearts and mix until incorporated.

Spoon the mixture into the hollowed-out loaf of bread. Wrap in aluminum foil and bake for 40 minutes.

Just before serving sprinkle the loaf with chopped parsley. Surround the loaf with the reserved pieces of bread for dipping.

ROASTED RED PEPPER DIP

1	small head garlic, top removed	1
3	sweet red peppers, roasted	3
2 drops	hot pepper sauce	2 drops
1	4-oz. (113-g) package cream cheese, softened	1
3 Tbsp.	sour cream	45 mL
	salt and freshly ground black pepper to taste	

Makes 1 1/2 cups (360 mL)

Preheat the oven to 375°F (190°C).

Drizzle the head of garlic with a little oil, wrap in aluminum foil and bake for approximately 40 minutes, until the garlic is soft. When it is cool enough to handle, squeeze the garlic cloves from their skins.

Place the roasted garlic and red peppers, hot sauce and cream cheese in a food processor and blend until smooth. Transfer to a bowl and stir in the sour cream. Season with salt and pepper.

Cover and refrigerate for 2 hours. This dip keeps well refrigerated for up to 3 days.

The red pepper and garlic in this dip provide a sweet nutty flavor that only roasting vegetables can give. Serve the dip with small pieces of bread, such as pita triangles.

ROASTING PEPPERS

Cut the peppers in half and discard the seeds. Brush the skin side of the peppers with vegetable oil. Place them under the broiler, skin side up, until the skins are blackened and blistering. (Or grill them skin side down.) When the skins have blackened, place them in a bowl and cover tightly with plastic wrap. When they're cool enough to handle, about 8 to 10 minutes, remove the skins and discard.

BLISSFUL DIP

1 cup	mayonnaise	240 mL
1 cup	sour cream	240 mL
2 tsp.	chopped fresh garlic	10 mL
1 tsp.	dried dill weed	5 mL
1 tsp.	celery seed	5 mL
1 tsp.	beaumond	5 mL
1 tsp.	dried parsley	5 mL
2 drops	hot pepper sauce	2 drops

*Makes 2 cups
(480 mL)*

This dip is even better made a day ahead. Serve it with crackers or an assortment of vegetables —raw or marinated, roasted or blanched— for dipping. Bread sticks, marinated artichoke hearts and marinated mushrooms all add interesting flavors and textures to the platter.

Combine all the ingredients in a bowl, mix well and chill for about 2 hours before serving. This dip keeps refrigerated for up to 3 days.

Note: Beaumond is a spice blend available at most grocery and bulk food stores.

Black Olive and Sun-Dried Tomato Tapenade

1/2 cup	sun-dried tomatoes, reconstituted	120 mL
1 cup	black oil-cured olives, pitted	240 mL
2 Tbsp.	capers, rinsed and drained	30 mL
2	cloves garlic	2
2 Tbsp.	lemon juice	30 mL
1/4 tsp.	freshly ground black pepper	1.2 mL
1/3 cup	extra virgin olive oil	80 mL
1	7-oz. (213-mL) can tuna, drained	1

Makes 2 cups
(480 mL)

In a food processor combine the sun-dried tomatoes, olives, capers, garlic, lemon juice and pepper. Process to form a paste. With the motor running, slowly add the olive oil. Transfer to a small bowl, add the tuna and mix well. Refrigerate for 2 hours before serving.

Tapenade can be stored in the refrigerator for up to 1 week.

A truly delightful savory spread, tapenade goes especially well with goat cheese on garlicky toasted rounds of bread. It also makes a tasty dip, served with a large platter of fresh vegetables.

Reconstituting Sun-Dried Tomatoes

Bring a saucepan of water to a boil. Add the sun-dried tomatoes and simmer, just covered with water, for 3 minutes, until tender.

GUACAMOLE

2	very ripe avocados	2
1/2	tomato, chopped	1/2
1 Tbsp.	lime juice	15 mL
1/4 cup	finely chopped onion	60 mL
1	clove garlic, finely minced	1
1/2	jalapeño pepper, seeded and finely chopped	1/2
3/4 tsp.	salt (or to taste)	4 mL

Serves 6 to 8

Use guacamole alongside Quesadillas (page 186) or as a topping for burgers and tacos.

Halve and pit the avocados and scoop the flesh into a bowl. Cut the avocado into 1/4-inch (.6-cm) cubes. Add the remaining ingredients and toss together. Taste and adjust the seasoning. Serve immediately.

HUMMUS

2 cups	canned chick peas, drained	480 mL
1/4 cup	tahini (sesame paste)	60 mL
2 Tbsp.	warm water	30 mL
2 Tbsp.	extra virgin olive oil	30 mL
1/4 cup	lemon juice	60 mL
2	cloves garlic, minced	2
1 tsp.	cumin seed	5 mL
1/2 tsp.	cayenne pepper	2.5 mL
	freshly ground black pepper to taste	
2 Tbsp.	chopped fresh parsley (optional)	30 mL

Makes 2 cups (480 mL)

A classic summertime dip with pita bread. Hummus can be stored in the refrigerator for up to 2 days.

Place all the ingredients except the parsley in a food processor and pulse until well combined. Transfer to a bowl. Garnish with chopped parsley, if desired.

BABA GHANNOUJ

1 lb.	eggplant	455 g
1	head garlic	1
1/4 cup	tahini (sesame paste)	60 mL
1/4 cup	lemon juice	60 mL
2 tsp.	salt	10 mL
1/2 tsp.	freshly ground black pepper	2.5 mL
2 Tbsp.	olive oil	30 mL
6 drops	hot pepper sauce, or to taste	6 drops
	pinch sugar (optional)	
2 Tbsp.	finely chopped fresh coriander	30 mL

Makes 1 1/2 cups
(360 mL)

A classic Middle Eastern dip made with baked eggplant and lots of garlic. Serve it with pita triangles the next time company's coming.

Preheat the oven to 400°F (200°C).

Prick the eggplant all over with a fork and place it on a baking sheet. Cut about 1/4 inch (.6 cm) from the top of the head of garlic, drizzle a little olive oil over it, wrap it in aluminum foil and place it on the baking sheet. Bake both for approximately 40 to 55 minutes, or until soft. Remove from the oven and let cool.

Cut the eggplant in half and scoop the pulp into a food processor. Squeeze the garlic cloves from their skins and add to the food processor. Add the tahini, lemon juice, salt, pepper, olive oil and hot pepper sauce. Blend the ingredients until the mixture is a smooth paste. Taste and add a few more drops of hot pepper sauce if you want a little more heat. If the mixture tastes a little bitter, add a pinch or two of sugar. Process a few seconds longer.

Place the mixture into a serving bowl. Mix in the coriander, reserving 1 tsp. (5 mL). Refrigerate for a few hours before serving. Garnish with the remaining chopped coriander just before serving.

Mushroom-Hazelnut Pâté with Cognac

*Makes about
1 1/2 cups (360 mL)*

*This is a great make-
ahead pâté recipe.
Refrigerate it overnight to
allow the flavors to ripen.*

2 Tbsp.	butter	30 mL
1/4 cup	shallots, chopped	60 mL
1/4 cup	finely shredded peeled carrot	60 mL
1 Tbsp.	cognac	15 mL
2 Tbsp.	butter	30 mL
1 lb.	fresh mushrooms, sliced (about 4 cups/1 L)	455 g
1/4 cup	hazelnuts, toasted (see page 183)	60 mL
2 Tbsp.	cognac	30 mL
	pinch ground allspice	
	pinch cayenne pepper	
	grating of fresh nutmeg	
	salt and freshly ground black pepper to taste	

Melt 2 Tbsp. (30 mL) of butter in a medium sauté pan over
medium-high heat. Add the shallots and sauté for 1 minute. Add
the carrot and sauté for 2 minutes more, or until the shallots are
soft. Add 1 Tbsp. (15 mL) of cognac and sauté until evaporated.
Transfer the mixture to a food processor.

Place the remaining 2 Tbsp. (30 mL) of butter in the pan, add the
mushrooms and sauté until soft, about 5 minutes. Transfer to the
food processor. Add the hazelnuts, remaining 2 Tbsp. (30 mL) of
cognac, allspice, cayenne pepper and nutmeg. Season with salt
and pepper. Purée until smooth.

Transfer the mixture to a pâté dish or small crock, cover and
refrigerate overnight. Serve with crackers.

HERBED SAVORY CHEESECAKE

12 oz.	cream cheese, softened	340 g
2	eggs	2
2 Tbsp.	whipping cream	30 mL
	freshly ground black pepper to taste	
	pinch cayenne pepper	
1	sweet red pepper, roasted	1
	(see page 5) and chopped	
1/4 cup	chopped fresh chives	60 mL
2 tsp.	chopped fresh dill	10 mL
1	small clove garlic, crushed	1
1 cup	red pepper jelly	240 mL

Serves 12 to 18

This rich cheesecake topped with pepper jelly has a little zing in every bite.

Butter a 7-inch (18-cm) springform pan. Line the bottom with a piece of parchment paper. To get the right size, place the pan on the parchment, draw the outline and cut out the circle. Place the circle in the springform pan.

Preheat the oven to 325°F (165°C).

Place the cream cheese, eggs, cream, black pepper and cayenne pepper in a food processor. Process until smooth. Transfer to a mixing bowl. Add the red pepper, chives, dill and garlic. Mix until well combined.

Pour the batter into the prepared pan. Bake for 30 minutes or until it's just beginning to brown. Remove from the oven and cool on a rack.

In a small saucepan, melt the red pepper jelly over medium heat, stirring often. Pour it over the cooled cheesecake. Refrigerate for several hours or overnight before serving.

Note: Place a baking sheet under the springform pan while baking the cheesecake, just in case the pan leaks.

Antipasto

**Makes approximately
12 cups (3 L)**

*Served with crackers, this
appetizer is great for a
large crowd. It keeps
well for 3 weeks in the
refrigerator, so make
a batch to have on
hand for the holiday
entertaining season.*

1/2	*small cauliflower*	*1/2*
2	*7 1/2-oz. (213-mL) cans chunk light tuna, drained*	*2*
2	*sweet green peppers, chopped into 1/4-inch (.6-cm) pieces*	*2*
1	*sweet red pepper, chopped into 1/4-inch (.6-cm) pieces*	*1*
1	*13-oz. (370-mL) jar stuffed green manzanilla olives*	*1*
1/2 cup	*hot pickled banana pepper, drained and finely chopped*	*120 mL*
1	*13-oz. (370-mL) jar sweet pickled onions*	*1*
2	*8-oz. (227-mL) cans whole mushrooms, drained*	*2*
2 cups	*tomato ketchup*	*480 mL*
1/2 cup	*sweet green relish*	*120 mL*
1/2 cup	*chili sauce*	*120 mL*
1/2 cup	*white vinegar*	*120 mL*
1/4 cup	*extra virgin olive oil*	*60 mL*
1 1/2 tsp.	*ground cinnamon*	*7.5 mL*
1/4 tsp.	*cayenne pepper*	*1.2 mL*

In a medium-size saucepan bring salted water to a boil. Break the cauliflower into medium-size pieces. Blanch in the boiling water for 3 to 5 minutes, or until just barely tender. Drain, plunge immediately into cold water, drain again and break into smaller pieces.

Place all the ingredients except the cauliflower in a large pot. Bring the mixture to a boil, reduce the heat to low and simmer for 10 minutes, stirring frequently. Add the cauliflower and simmer 5 minutes longer.

Remove from the heat and cool for 10 minutes. Store in a glass container in the refrigerator for 3 to 5 days before serving to allow the flavors to meld.

MARINATED MUSHROOMS

1 lb.	white mushrooms	455 g
1 1/2 cups	white wine vinegar	360 mL
1 1/2 cups	water	360 mL
6	black peppercorns	6
2	whole cloves	2
1/2 tsp.	mustard seed	2.5 mL
1	clove garlic	1
	few leaves rosemary, bruised to release the oils	
1	bay leaf	1
1	1/2-inch (1.2-cm) piece cinnamon stick	1
3 Tbsp.	extra virgin olive oil	45 mL
1/4 tsp.	dried basil	1.2 mL
1/4 tsp.	dried oregano	1.2 mL
1/4 tsp.	dried thyme	1.2 mL
1/4 tsp.	dried parsley	1.2 mL
	salt and freshly ground black pepper to taste	

*Makes approximately
1 cup (240 mL)*

*These mushrooms are
perfect for an antipasto
platter and a great
addition to any salad.*

Choose smaller mushrooms that are uniform in size. Wash them and trim the stems.

Combine the vinegar and water in a small pot and bring the mixture to a boil. Place the peppercorns, cloves, mustard seed, garlic, rosemary, bay leaf and cinnamon stick in cheesecloth and tie it with string. Place the spice bag and mushrooms in the vinegar mixture. Return it to a boil, reduce the heat and simmer for 10 minutes. Turn off the heat and let the mixture sit for 2 hours.

Drain the mushrooms and discard the spice bag. Rinse the mushrooms and dry on paper towel.

Combine the olive oil, basil, oregano, thyme, parsley, salt and pepper. Drizzle over the mushrooms, toss and refrigerate until serving time.

The mushrooms will keep for 3 days refrigerated in an airtight container.

CHERRY TOMATOES FILLED WITH SUN-DRIED TOMATOES

40	*red or yellow cherry tomatoes*	40
2 Tbsp.	*sun-dried tomatoes*	30 mL
4 oz.	*cream cheese, softened*	113 g
3 1/2 oz.	*chèvre cheese*	100 g
2 Tbsp.	*olive oil*	30 mL
1 Tbsp.	*finely chopped fresh basil*	15 mL
1 tsp.	*finely grated lemon zest*	5 mL
10	*whole basil leaves*	10

Makes 40

Perfect for the tomato lover and will make a convert of those who aren't. Use the sweet little tomatoes you get right off the vine in summer. If you're able to find yellow tomatoes, try them for a different look.

Wash and dry the cherry tomatoes. Remove the bottom 1/4 of each tomato with a sharp knife. With a small melon-ball scoop, remove the seeds. Place the tomatoes on paper towel, cut side down, to drain. Refrigerate until you're ready to stuff them.

Bring a small saucepan of water to a boil and simmer the sun-dried tomatoes for about 3 minutes, just until tender. Drain and chop.

Place the sun-dried tomatoes, cream cheese, chèvre, oil, basil and lemon zest in a food processor and process until very smooth.

Stuff the drained tomatoes with the cheese filling, using a pastry bag fitted with a small tip. Garnish with a thin slice of basil leaf.

BANANA PEPPERS STUFFED WITH CHÈVRE AND FETA

6	sweet banana peppers	6
2 Tbsp.	olive oil	30 mL
2 Tbsp.	chopped green onion	30 mL
2 Tbsp.	chopped walnuts, toasted (see page 183)	30 mL
3 1/2 oz.	chèvre, crumbled	100 g
1/4 cup	feta cheese, crumbled	60 mL
1 Tbsp.	chopped fresh basil	15 mL
1 Tbsp.	chopped fresh parsley	15 mL
	few drops hot pepper sauce	
	freshly ground black pepper to taste	

Makes 30

This recipe calls for sweet banana peppers, but if you're feeling a little daring, try using hot ones.

Heat the barbecue to medium hot.

Brush the peppers with the olive oil and grill until the skin is blistered and charred. Place the peppers in a bowl and cover with plastic wrap. When they're cool enough to handle, peel off the skins. Remove the stems and make a slit along the length of each pepper. Open the pepper and remove the seeds. Set the prepared peppers aside.

Place the green onion, walnuts, goat and feta cheeses, basil, parsley, hot sauce and pepper in a food processor. Process just until well blended.

Divide the cheese filling evenly among the peppers. Close the peppers around the filling, keeping their original shape. Chill them for at least 2 hours, or overnight. Cut each pepper crosswise into 5 slices.

Walnut and Cheddar Stuffed Mushrooms

Makes approximately 16

Walnuts and Cheddar cheese team up in this classic. A must for mushroom lovers.

1 lb.	large fresh mushrooms	455 g
6 Tbsp.	butter, melted	90 mL
1 cup	finely chopped onion	240 mL
1 cup	soft bread crumbs	240 mL
1 cup	Cheddar cheese, grated	240 mL
1/2 cup	walnuts, toasted (see page 183) and finely chopped	120 mL
1/4 cup	finely chopped fresh parsley	60 mL
	salt and freshly ground black pepper to taste	

Preheat the oven to 350°F (175°C).

Clean the mushrooms. Remove the stems and chop finely. Place the caps on a baking sheet.

Brush the caps with some of the melted butter. Heat the remaining butter in a sauté pan over medium-high heat. Add the onions and mushroom stems. Sauté until the onions are soft and translucent. Remove from the heat, add the bread crumbs, Cheddar cheese, walnuts and parsley. Season with salt and pepper. Stir the mixture until it's well combined.

Divide the filling evenly among the mushroom caps, mounding the filling high in each one. Bake for 20 minutes. Serve hot.

CUCUMBER AND RADISH ROUNDS WITH CHÈVRE

7 oz.	chèvre cheese	200 g
1 Tbsp.	finely chopped shallot	15 mL
1/4 cup	finely chopped chives	60 mL
2 Tbsp.	finely chopped fresh parsley	30 mL
1 Tbsp.	finely chopped fresh chervil	15 mL
1/3 cup	whipping cream, whipped into soft peaks	80 mL
2 Tbsp.	olive oil	30 mL
1 1/2 tsp.	white wine vinegar	7.5 mL
	salt and freshly ground black pepper to taste	
12	radishes, stems attached	12
1	English cucumber	1

*Makes 2 cups
(480 mL)
cheese spread and
approximately
48 canapés*

Using an electric mixer, beat the cheese with the shallot, chives, parsley and chervil. Fold in the whipped cream, olive oil, vinegar, salt and pepper. Chill well.

Clean and trim the radishes, leaving 1/2 inch (1.2 cm) of stem attached. Slice each radish in half, making sure a little stem remains on each half. Remove a small slice from the round side of the radish so it will lie flat.

Slice the cucumber into 1/4-inch (.6-cm) slices.

To assemble, place the cheese mixture into a pastry bag fitted with a medium tip. Pipe the mixture onto the cucumber and radish slices. Arrange on a platter and garnish with fresh herbs.

Chèvre is the star in this recipe. We've served the cheese spread on radish and cucumber slices, but it's just as good spread on crackers or as a dip for fresh vegetables.

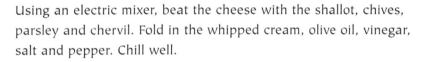

MARINATED TRIO OF ROASTED PEPPERS

Makes about 2 cups (480 mL)

Add the peppers to pizza or serve them on grilled baguettes with goat cheese. They look particularly nice in a bowl with marinated and spiced olives.

4	cloves garlic, peeled and halved	4
1 Tbsp.	chopped assorted fresh herbs, such as thyme, oregano, chives and basil	15 mL
1/4 cup	white wine vinegar	60 mL
3/4 cup	olive oil	180 mL
2 lbs.	mixed sweet orange, yellow and red peppers, roasted (see page 5)	900 g

Preheat the grill to medium-high. Cut each pepper half into quarters.

In a 1-quart (1-L) glass jar, combine the garlic, herbs, wine vinegar and olive oil. Add the peppers. Marinate overnight in the refrigerator before using.

The peppers will keep for up to 1 week refrigerated.

Potato Galettes with Crème Fraîche and Smoked Trout

3	medium red-skinned potatoes, scrubbed well	3
	salt and freshly ground black pepper to taste	
1 cup	Crème Fraîche (see page 214)	240 mL
8 oz.	smoked trout	225 g

Preheat the oven to 400°F (200°C).

Line 2 baking sheets with parchment paper.

Using a mandoline or very sharp knife, slice the potatoes very thinly—1/8 inch (.3 cm). Brush each slice with vegetable oil. Arrange the potato slices in groups of 3 on the baking sheets. Overlap the 3 slices in the center, making a shamrock shape. Season with salt and pepper. Place in the oven and bake 6 to 8 minutes, or just until the potatoes are cooked.

Remove from the oven, dot with a little crème fraîche, add a little smoked trout and serve on a warmed plate.

Makes approximately 30

You can add any topping to these soft potato cakes—caviar, salmon, smoked meats—the possibilities are endless.

ROASTED GARLIC WITH BRIE BRUSCHETTA

Makes approximately 16

Roasting garlic gives it a much milder flavor, which pairs perfectly with the Brie in this recipe.

4	heads fresh garlic	4
	salt and freshly ground black pepper to taste	
	olive oil	
1	baguette	1
1/2 lb.	Brie cheese, thinly sliced	225 g
1	sweet red pepper, roasted (see page 5) and cut into long strips	1
1/4 cup	sliced black olives	60 mL

Preheat the oven to 375°F (190°C).

Remove the top 1/4 inch (.6 cm) of each head of garlic. Sprinkle with salt and freshly ground black pepper, drizzle with a little olive oil and wrap in aluminum foil. Bake for 40 minutes, or until soft. Remove from the oven. Squeeze the garlic cloves into a bowl. Mash the garlic with a fork until it becomes a spreadable paste. Set aside.

Cut the baguette into 1-inch (2.5-cm) slices. Brush the slices lightly with olive oil, sprinkle with salt and pepper, place on a baking sheet and broil until lightly toasted. Turn and lightly toast the other side.

Spread a little of the mashed garlic onto each slice of toasted baguette. Cover with a slice of Brie and make an X across the top of the cheese with thin slices of red pepper. Garnish with a few black olive slices. Return to the oven and broil until the cheese begins to melt. Serve immediately.

CARAMELIZED ONION AND GORGONZOLA CROSTINI

1	baguette	1
1	clove garlic, halved	1
1/4 cup	olive oil	60 mL
2 lbs.	onions, thinly sliced	900 g
	pinch sugar	
1	bay leaf	1
1/2 cup	red wine	120 mL
1 Tbsp.	chopped fresh thyme	15 mL
1/4 lb.	Gorgonzola cheese, crumbled	113 g

Makes approximately 16

Caramelizing onions brings out a sweetness that's hard to resist; pairing them with Gorgonzola cheese is practically sinful.

Preheat the broiler.

Slice the baguette into 1-inch (2.5-cm) slices on the diagonal. You will have approximately 16 pieces. Rub each piece with the cut side of the garlic clove. Brush lightly with some of the olive oil. Place on a baking sheet and broil until lightly browned. Turn and lightly brown the other side.

Remove from the oven and set aside. Heat the remaining olive oil in a large skillet over medium-low heat. Add the onions, sugar and bay leaf. Cook until the onions are soft and brown, stirring occasionally. It will take approximately 30 minutes. Add the red wine and thyme and continue to cook until all the wine has evaporated. Discard the bay leaf.

Distribute the filling evenly over the toasted baguette slices. Sprinkle a little Gorgonzola over each piece. Place under the broiler until the cheese has melted.

Note: For a quick appetizer, prepare the baguette and onion a day ahead. Store the crostini in an airtight container and the onion in the refrigerator. Be sure to reheat the onions before assembling.

TOMATO BASIL BRUSCHETTA

Makes approximately 16

Bruschetta makes a great opener for an Italian meal. These toasted baguettes have a wonderful tomato topping.

1	baguette	1
6	plum tomatoes, finely chopped	6
1/4 cup	chopped fresh basil	60 mL
2	cloves garlic, finely chopped	2
1/4 cup	extra virgin olive oil	60 mL
1/2 cup	grated Parmesan cheese	120 mL
	salt and freshly ground black pepper to taste	

Cut the baguette into 1-inch (2.5-cm) slices on the diagonal. You will have approximately 16 pieces. Brush each piece lightly with some of the olive oil. Place on a baking sheet and broil until lightly browned. Turn and lightly brown the other side. Remove from the oven and set aside.

In a large bowl mix together the tomatoes, basil, garlic, olive oil and 1/4 cup (60 mL) of the Parmesan cheese. Mix well. Season with salt and pepper.

Spoon the tomato mixture over the baguette slices and sprinkle the remainder of the Parmesan cheese on top. Place under the broiler until just warmed.

Serve immediately.

Note: The baguette can be prepared a day ahead and stored in an airtight container. The tomato mixture can be made a few hours ahead of serving.

Sun-Dried Tomato Palmiers

1 cup	sun-dried tomatoes, reconstituted (see page 7)	240 mL
3/4 cup	fresh basil	180 mL
1/2 cup	pitted black olives, drained	120 mL
1/2 cup	pine nuts	120 mL
1 Tbsp.	anchovy paste	15 mL
2	cloves garlic	2
1/4 cup	olive oil	60 mL
2	16-oz. (455-g) packages frozen puff pastry, thawed	2
1	egg, lightly beaten	1

Makes 48

These little bites just melt in your mouth. They can be stored in the fridge for up to 2 days or frozen for up to 2 weeks. Reheat to serve.

In a food processor blend the tomatoes, basil, olives, pine nuts, anchovy paste and garlic until finely chopped. With the motor running, gradually add the olive oil.

Divide the puff pastry into 4 equal pieces.

Roll each piece out to a 6- x 14-inch (15- x 36-cm) rectangle. Spread 1/3 cup (80 mL) of filling on each rectangle, spreading it to within 1 inch (2.5 cm) of the edge of the pastry. Roll both long sides of the pastry toward the center, making a heart shape. Brush the beaten egg mixture along the center where the two sides meet and press them together. Cover with waxed paper and refrigerate for 30 minutes. Repeat this procedure with the other 3 rectangles.

Preheat the oven to 375°F (190°C).

Cut the logs into 3/4-inch (1.9 cm) slices. Place on a lightly oiled baking sheet. Bake 12 to 15 minutes, or until the palmiers are golden brown. Serve hot.

ZUCCHINI HERB BLINI WITH SMOKED SALMON

Makes 40

We've topped these miniature pancakes with smoked salmon for an elegant hors d'oeuvre. Blini can be made ahead or even frozen for up to 2 weeks.

1 cup	all purpose flour	240 mL
1 1/2 tsp.	baking powder	7.5 mL
1 tsp.	baking soda	5 mL
1/8 tsp.	salt	.5 mL
3	eggs, separated	3
1/4 cup	milk	60 mL
1 3/4 cups	sour cream	420 mL
1/4 cup	finely grated zucchini	60 mL
4 Tbsp.	chopped fresh herbs, any mixture of thyme, basil, savory, sage or chives	60 mL
1/2 tsp.	crushed red chili peppers	2.5 mL
1/2 cup	sour cream	120 mL
1 Tbsp.	bottled horseradish	15 mL
2 Tbsp.	chopped fresh herbs	30 mL
8 oz.	smoked salmon	225 g
40	fresh herb sprigs	40

Sift the flour, baking powder, baking soda and salt together in a large bowl.

Beat the egg whites until stiff but not dry.

Combine the egg yolks, milk, the 1 3/4 cups (420 mL) of sour cream and the grated zucchini. Stir until well blended. Add the flour mixture to the egg mixture with a few quick strokes. Stir in 2 Tbsp. (30 mL) of the fresh herbs and the chili pepper. Fold in the egg whites. Do not overmix.

Heat a lightly buttered griddle to medium-hot. Drop the batter by spoonfuls onto the griddle. The pancakes should be 2 to 3 inches (5 to 7.5 cm) in diameter. When bubbles appear on the surface of the cakes, turn them over and cook 2 minutes more, or until golden brown on the other side.

Let the pancakes cool to room temperature. Do not stack them on top of each other until they are completely cooled, or they will become soggy. Blinis can be refrigerated for up to 2 days or frozen for 2 weeks. Thaw and serve them at room temperature.

Combine the remaining 1/2 cup (120 mL) of sour cream, the horseradish and the remaining 2 Tbsp. (30 mL) of chopped herbs.

Spoon a little sour cream mixture on top of each blini. Top with some smoked salmon and garnish with a fresh herb sprig.

Snow Peas Stuffed with Stilton and Walnuts

36	*snow peas*	*36*
1/4 cup	*walnuts, toasted (see page 183)*	*60 mL*
4 oz.	*Stilton cheese*	*113 g*
4 oz.	*cream cheese, softened*	*113 g*
2 Tbsp.	*whipping cream*	*30 mL*

Makes 36

Remove the stem end and the string from the snow peas. Bring a pot of water to a boil. Blanch the peas for 30 seconds, remove them with a slotted spoon and plunge them immediately into cold water to halt the cooking process. Drain on paper towel.

Grind the walnuts in a food processor until very fine. Add the Stilton, cream cheese and cream. Process just until combined.

With a knife tip, cut the snow peas open lengthwise. Using a pastry bag fitted with a small tip, pipe the filling into each snow pea.

This is finger food at its most prestigious. It's a little fussy to make but well worth the effort.

RED PEPPER AND CHUTNEY PHYLLO TRIANGLES

6 oz.	*chèvre cheese*	170 g
1/4 cup	*finely chopped sweet red pepper*	60 mL
1/4 cup	*finely chopped green onion*	60 mL
1 tsp.	*mild curry powder*	5 mL
1	*8-oz. (227-mL) jar mango chutney*	1
1	*16-oz. (455-g) package frozen phyllo, thawed*	1
1/2 cup	*butter*	120 mL

Makes approximately 24

A surprisingly sweet, hot filling in every bite. Pass around the remaining chutney for dipping these delicious triangles.

In a food processor, mix the chèvre, red pepper, green onion, curry powder and 2 Tbsp. (30 mL) of the chutney. Process until just combined.

Melt the butter in a small saucepan. Add 2 Tbsp. (30 mL) chutney to the butter. Keep the mixture warm.

Preheat the oven to 425°F (220°C).

Unwrap the phyllo and place one sheet on your work surface. (Cover the remaining phyllo with wax paper and a damp cloth to prevent it from drying out.) Brush the phyllo sheet with the melted butter mixture. Place a second sheet on top. Brush with more butter and repeat with a third sheet.

With a sharp knife, cut the phyllo into 6 strips. Place 1 tsp. (5 mL) of the cheese mixture and 1/4 tsp. (1.2 mL) of chutney about an inch (2.5 cm) from the edge nearest to you. Pick up the lower right corner and fold it diagonally over the filling, making a triangle. Now pick up the bottom of the triangle and fold it up. Continue folding in this fashion until you reach the end of the strip. (If you've never made phyllo triangles, practice on a piece of paper first. You must work fast to ensure that the phyllo doesn't dry out.) Brush the tops of the triangles with the butter and chutney mixture. Place on parchment-lined baking sheets.

Repeat with the remaining filling and phyllo. Return the unused portion of the phyllo to the freezer.

Bake the triangles 7 to 8 minutes, or until golden brown.

Note: The triangles can be frozen unbaked for up to a month in airtight containers. Do not thaw before baking, but increase the baking time to approximately 10 minutes.

INDIAN VEGETABLE PAKORAS

1 cup	all purpose flour	240 mL
	pinch baking soda	
1/2 tsp.	cayenne pepper	2.5 mL
1/2 tsp.	salt	2.5 mL
3/4 cup	water	180 mL
	vegetable oil for deep frying	
2	medium red potatoes, scrubbed and cut into small cubes	2
1 cup	peas, fresh or frozen	240 mL
1 cup	carrots, peeled and cut into cubes	240 mL
18	spinach leaves, washed, dried and thinly sliced	18
16	small cauliflower florets	16

Makes 24

These spicy vegetable fritters hail from India. For an authentic pakora, substitute chick pea flour for all purpose flour if it is available. Serve the pakoras with a few different chutneys as an appetizer or as a side dish.

In a small bowl mix the flour, baking soda, cayenne pepper and salt. Add the water and stir until a smooth batter is formed.

Heat the oil over medium-high heat. It's ready when a little batter rises quickly to the top.

Combine the batter with the vegetables, and drop by spoonfuls into the oil. Fry the fritters until they are golden brown. Drain on paper towel and keep warm until serving time.

Savory Pepper and Mushroom Profiteroles

Makes about 48

Sweet peppers and mushrooms sautéed and presented in tiny choux pastry casings—these are worth the effort.

For the choux pastry:

1 cup	water	240 mL
1/2 cup	unsalted butter, cut into small pieces	120 mL
1/2 tsp.	salt	5 mL
1 cup	all purpose flour	240 mL
4	eggs	4

Bring the water, butter and salt to a simmer in a medium saucepan over medium heat. When the butter has melted, reduce the heat to low and whisk in the flour. Beat with a wooden spoon until the batter is smooth and shiny and leaves the sides of the pan clean. Remove from the heat.

Beat in the eggs one at a time, taking care to thoroughly incorporate each one into the batter before adding the next.

Preheat the oven to 450°F (230°C).

Line 2 baking sheets with parchment paper. Using a pastry bag fitted with a plain round tip, pipe the batter onto the sheets in 1-inch (2.5-cm) mounds.

Bake for 15 minutes, reduce the heat to 300°F (150°C) and bake about 20 minutes longer, until the puffs are golden brown and very light. Place on a rack to cool.

For the filling:

1 Tbsp.	vegetable oil	15 mL
1 Tbsp.	butter	15 mL
1	medium onion, finely chopped	1
4	sweet bell peppers, assorted colors, julienned	4
1/2	jalapeño pepper, seeded and finely chopped	1/2
1/2 lb.	mushrooms, sliced	225 g
2	cloves garlic, finely chopped	2
	salt and freshly ground black pepper to taste	
2 Tbsp.	balsamic vinegar	30 mL
2 Tbsp.	finely chopped fresh parsley	30 mL
2 Tbsp.	finely sliced green onions	30 mL

Heat the oil and butter over medium-high heat in a large skillet. Add the onion and sauté until it starts to soften. Add the sweet and hot peppers and mushrooms. Cook until the peppers begin to soften. Reduce the heat, stir in the garlic and cook another 10 minutes, until the vegetables are tender and starting to turn a light golden color.

Season with salt and pepper and increase the heat to high. Add the balsamic vinegar and cook until the vinegar is almost evaporated, stirring often. Remove from the heat and stir in the parsley and green onions, reserving some of the parsley for garnishing.

To assemble:

Slice the top 1/3 off the profiteroles and hollow out the contents. Stuff with the sautéed vegetables.

Sprinkle the reserved chopped parsley over the top as a garnish and serve warm.

WILD MUSHROOM PHYLLO PURSES

For the phyllo purses:

2 Tbsp.	butter	30 mL
2	shallots, finely chopped	2
1 lb.	assorted wild mushrooms, coarsely chopped	455 g
1/4 cup	white wine	60 mL
1 tsp.	fresh thyme	5 mL
1 tsp.	chopped fresh rosemary	5 mL
1/2 cup	whipping cream	120 mL
1	sweet red pepper, roasted (see page 5) and coarsely chopped	1
12 oz.	Gruyère cheese, grated salt and freshly ground black pepper to taste	340 g
12	sheets frozen phyllo, thawed	12
1/4 cup	butter, melted	60 mL

Serves 8

This recipe works just as well with domestic mushrooms.

Melt the 2 Tbsp. (30 mL) butter in a large sauté pan over medium-high heat. Add the shallots and sauté for 2 minutes. Add the mushrooms and sauté until all the liquid evaporates. Add the white wine and simmer until the liquid is reduced by half. Add the thyme, rosemary and cream, and bring to a boil. Remove from the heat. Stir in the red pepper and grated cheese. Season with salt and pepper.

Lay one sheet of phyllo on your work surface and brush with melted butter. Place a second sheet on top of the first one, brush it with melted butter, and repeat with a third sheet. With a sharp knife cut the sheet into 4 rectangles.

Place a large tablespoon of mushroom filling in the center of each rectangle. Pick up the corners of the first rectangle, bring them into the center above the filling and give a gentle twist to close the purse. Brush the top with butter and place on a non-stick or parchment-lined baking sheet. Continue with the next 3 rectangles. Repeat with the remaining phyllo sheets.

Preheat the oven to 400°F (200°C).

Bake the phyllo purses for 15 minutes, or until golden brown. Check regularly to be sure they don't brown too fast.

For the sauce:

1 lb.	*domestic mushrooms*	*455 g*
2	*shallots*	*2*
2 Tbsp.	*butter*	*30 mL*
	salt and freshly ground black pepper	
	to taste	
1/4 cup	*white wine*	*60 mL*
1 cup	*whipping cream*	*240 mL*
12	*whole chive stalks*	*12*

In a food processor, finely chop the mushrooms. Remove the mushrooms and finely chop the shallots. Melt the butter in a sauté pan over medium heat. Add the shallots and sauté lightly for 1 minute. Add the mushrooms and season with salt and pepper. Cook until all the liquid has evaporated, add the white wine and simmer until the liquid is reduced by half. Add the whipping cream and adjust the seasoning.

Blanch the 12 chive stalks in boiling water for a few seconds to make them soft. Plunge them into cold water immediately. Drain and dry them on a paper towel.

To serve, tie a softened chive stalk around each purse and trim it neatly. Pour a small pool of sauce onto each plate. Place 2 purses on each plate. Serve warm.

FIERY VEGETABLE SAMOSAS

Makes 24 samosas

We've called for Thai spring roll wrappers to make these samosas but if you're feeling adventurous, try using phyllo pastry. The spring roll wrappers are available in the frozen food section at Asian grocery stores.

3 Tbsp.	vegetable oil	45 mL
1 tsp.	cumin seed	5 mL
1/2 tsp.	coriander seed	2.5 mL
1	medium onion, finely chopped	1
2 tsp.	minced fresh ginger	10 mL
1/2	jalapeño pepper, seeded and chopped	1/2
1 Tbsp.	ground coriander	15 mL
1/2 tsp.	cayenne pepper	2.5 mL
3	large potatoes, peeled and boiled	3
1/2 cup	cooked green peas or frozen peas, thawed	120 mL
1/2 tsp.	salt	2.5 mL
1 Tbsp.	finely chopped fresh cilantro	15 mL
1/4 cup	flour	60 mL
3 Tbsp.	water	45 mL
	vegetable oil for deep frying	
24	6- x 6-inch (12.5- x 12.5-cm) Thai spring roll wrappers	24
1	8-oz. (227-mL) jar chutney	1

Heat the oil in a medium sauté pan over medium-high heat. Add the cumin and coriander seeds and sauté for a few seconds. Add the onion and cook until it is browned, 8 to 12 minutes. Add the ginger and jalapeño pepper, sauté for 1 minute, and stir in the coriander and cayenne pepper. Cook and stir for 1 minute, until all the spices are well blended. Cool.

Dice the potatoes and place in a mixing bowl. Add the peas, salt, cilantro and the onion mixture. Mix well but not until it's mushy; the filling should be lumpy.

Make a paste with the flour and water. Heat the vegetable oil.

Working with one wrapper at a time, place one point of the wrapper toward you. Place 2 Tbsp. (30 mL) of filling in the center of the wrapper. Tuck in the sides, fold the top down over the filling, and brush a little of the flour and water mixture over the corner that's not folded. Roll up the samosa and press gently to make sure the corner with the flour and water paste adheres to the wrapper.

Deep-fry in batches, turning so that both sides are lightly browned. Drain on a paper towel and keep warm.

Serve with a little chutney for dipping.

Note: Wrap any unused spring roll wrappers in plastic wrap and refreeze for later use.

SPRING ROLLS WITH
THAI DIPPING SAUCE

Serves 6

These little bundles are just bursting with flavor. Have all the ingredients ready and roll them just before serving.

1/2 lb.	boneless chicken breasts	225 g
1 cup	bean sprouts	240 mL
1 cup	grated carrot	240 mL
3 Tbsp.	rice wine vinegar	45 mL
2 Tbsp.	sugar	30 mL
1/2 tsp.	salt	2.5 mL
1/2 tsp.	black pepper	2.5 mL
1 tsp.	vegetable oil	5 mL
5	egg yolks, beaten	5
6	rice paper sheets, 12 inches (30 cm) in diameter	6
1/4 cup	fresh coriander leaves	60 mL

Poach the chicken breasts in 2 to 3 cups (480 to 720 mL) of boiling water for 7 to 8 minutes. Do not overcook. Drain, cool and tear the chicken into thin strands by hand.

Bring 2 to 3 cups (480 to 720 mL) of water to a boil. Add the bean sprouts and cook for 30 seconds. Drain immediately. They will darken but will still be crunchy.

Place the grated carrot in a small bowl. Add the vinegar, sugar, salt and pepper. Mix well and set aside.

Heat the vegetable oil in a non-stick frying pan until medium hot. Add the egg yolks and cook until just firm (not too dry). They should not brown. Without turning the eggs, slide them from the pan. Roll up the omelette and cut it into thin slices. Set aside.

To assemble the spring rolls, fill a large bowl with very warm water. Place one rice paper sheet in the water. Carefully move it around until it becomes soft and pliable. Lay the sheet on a smooth work surface and flatten it out.

Onto the flattened rice paper place 1/6 of the chicken. Lay it evenly across the rice paper, leaving about 1 inch (2.5 cm) at the top edge of the wrapper and 1 inch (2.5 cm) at each side. On top of the chicken, place 1/6 of the bean sprouts, 1/6 of the carrot mixture, 1/6 of the omelette and 1/6 of the fresh chopped coriander. Fold the top edge over the filling, tuck in the sides and roll it up like a cigar. Repeat with the remaining 5 wrappers.

Slice each roll in half on the diagonal. Place one on top of the other. Serve with Thai dipping sauce.

Note: For a buffet service, cut the rolls into small pieces and arrange them on a platter.

THAI DIPPING SAUCE

Makes 1/2 cup (120 mL)

3 Tbsp.	Thai fish sauce	45 mL
3 Tbsp.	lime juice	45 mL
1 Tbsp.	chopped fresh garlic	15 mL
1 Tbsp.	Thai sweet chili sauce	15 mL
1 tsp.	finely diced jalapeño pepper	5 mL

Mix all ingredients in a bowl. Serve with Thai spring rolls.

MUSHROOMS IN PUFF PASTRY

Serves 8

Elegant and quick best describes this lovely mushroom appetizer.

1	1-lb. (455-g) package frozen puff pastry, thawed	1
1	egg, beaten	1
1 Tbsp.	butter	15 mL
2	shallots, finely chopped	2
2 cups	whipping cream	480 mL
1/2 cup	dry white wine	120 mL
1 Tbsp.	butter	15 mL
1 lb.	white or wild mushrooms, sliced	455 g
	salt and freshly ground black pepper to taste	

Preheat the oven to 400°F (200°C).

Working with one sheet at a time, roll out the dough on a lightly floured surface until it measures 10 x 10 inches (25 x 25 cm). Cut the sheet into quarters. Score the top of each piece with a sharp knife. Repeat with the second sheet.

Transfer the pastry to a heavy baking sheet. Brush the tops with the beaten egg. Bake the pastries until they're a deep golden color, approximately 10 minutes. Remove from the oven and cool.

Melt 1 Tbsp. (15 mL) of butter in a small saucepan over medium heat. Add the shallots and sauté until translucent, 1 to 2 minutes. Do not brown. Add the cream and white wine, bring to a boil and continue to cook until the sauce is reduced by half.

Meanwhile, in a medium sauté pan over medium heat, melt the remaining tablespoon (15 mL) of butter. Add the mushrooms and sauté until they begin to brown. Season with salt and pepper.

Remove the top of each puff pastry. Spoon the mushrooms on the bottom half and spoon the sauce over the mushrooms, dividing them equally among the 8 portions. Place the top of the pastry over the mushroom filling and serve immediately.

SOUPS

*What could be more welcoming to a guest
than a bowl of homemade soup?*

*Whether it's the main course for a casual après-ski
get-together or the starter to an elegant dinner party,
nothing is received with greater enthusiasm
than a delicious bowl of soup.*

CREAM OF ASPARAGUS SOUP

1 lb.	fresh green asparagus	455 g
1 Tbsp.	butter	15 mL
1/4 cup	chopped onion	60 mL
1/2 cup	chopped celery	120 mL
6 cups	chicken or vegetable stock	1.5 L
3 Tbsp.	butter	45 mL
3 Tbsp.	flour	45 mL
1 cup	whipping cream	240 mL
	salt and freshly ground black pepper to taste	
1/2 cup	sour cream	120 mL
	pinch paprika	

Serves 4

There's nothing quite as satisfying as a velvety smooth cream soup. Here's a way of using up asparagus stalks when you need the tips for a dish that's a little showier. Be sure to reserve a few of the tips to garnish the soup.

Wash and remove the tips from the asparagus. In a small saucepan bring 1/2 cup (120 mL) water to a boil. Add the tips and simmer for 2 minutes, until they are barely tender. Rinse under cold water to stop the cooking process. Drain and set them aside.

Cut the asparagus stalks into 1-inch (2.5-cm) pieces.

Melt the 1 Tbsp. (15 mL) of butter in a large saucepan over medium heat. Add the onion and celery and sauté until the onion is translucent, about 2 to 3 minutes. Add the asparagus stalks and the stock and bring the mixture to a boil. Reduce the heat to simmer and continue to cook for 30 minutes. Remove any scum that comes to the surface.

Remove the mixture from the heat and strain it through a sieve, pressing the vegetables with a wooden spoon to get all the liquid. Reserve the stock and discard the vegetables.

Melt the 3 Tbsp. (45 mL) of butter in a large saucepan over medium-high heat. Add the flour and stir until blended. Slowly add the cream, stirring constantly. Reduce the heat to low, add the stock and continue to stir until the soup has thickened. Do not boil it. Just before serving, add most of the reserved asparagus tips to heat them through. Season with salt and pepper.

Garnish each bowl with some sour cream, one or two asparagus tips and a sprinkle of paprika.

TIPS FOR MAKING THE PERFECT POT OF SOUP

- *Never sauté onions or leeks until brown unless it's specifically asked for in the recipe. Sauté until they are translucent and an iridescent yellow color. Sweat is the proper term. This technique makes the vegetables sweet, not bitter.*

- *Garlic should only be sautéed until fragrant; browning makes it very bitter. Usually 30 seconds to a minute in a moderately heated pan will do. Garlic is sautéed after the onions and just before adding the liquid; that way it will not be overcooked.*

- *Always remove any scum that forms on the surface of the soup; this prevents the soup from being bitter.*

- *Taste the soup before adding salt and pepper at the end of cooking. It's sometimes difficult to tell if more salt is needed, so use it sparingly. It's always easy to add more but pretty much impossible to remove.*

- *For velvety fine soups, purée in an electric blender. A food processor cannot get a soup quite as fine. Then pour it through a fine colander or chinois into a clean pot. You can use the same pot—just rinse it to get rid of any bits that may still be in the pot. Finish with a little heavy cream and butter, which gives the soup more body.*

- *For a lighter version, you can use half-and-half, milk or, in some cases, buttermilk. If you use these, do not boil the soup after adding them; only whipping cream can be boiled with any guarantee of not separating. Omit the butter completely.*

BORSCHT

Serves 6 to 8

Borscht always makes a colorful and delicious addition to the dinner table. Serve it with thick slices of pumpernickel bread and pass around a bowl of sour cream.

4 cups	peeled and grated beets	1 L
1/3 cup	red wine vinegar	80 mL
1 tsp.	sugar	5 mL
1/2 tsp.	caraway seeds	2.5 mL
1/4 cup	butter	60 mL
2 cups	finely chopped onion	480 mL
1 cup	coarsely grated carrot	240 mL
1	apple, peeled and grated	1
1	clove garlic, finely chopped	1
6 cups	beef stock	1.5 L
1 cup	coarsely shredded cabbage	240 mL
	salt and freshly ground black pepper to taste	
1 cup	sour cream	240 mL

Mix the grated beets with the vinegar, sugar and caraway seeds in a non-reactive bowl, such as glass. Set aside while preparing the rest of the vegetables.

Melt the butter over medium-high heat in a large stockpot. Add the onion and carrot and sauté until the onion begins to soften. Add the apple and garlic; sauté for 30 seconds. Add the beef stock and beet mixture. Bring the mixture to a boil.

Add the cabbage and continue to simmer until all the vegetables are tender, approximately 45 minutes. If the soup is too thick, add a little more beef stock. Adjust the seasoning with salt and pepper.

Garnish with large dollops of sour cream.

Cream of Beet, Potato and Leek Soup

4	medium beets, tops and roots trimmed to 1 inch (2.5 cm)	4
2 Tbsp.	butter	30 mL
2	leeks, white part only, washed and finely chopped	2
1	onion, finely chopped	1
4	potatoes, peeled and finely sliced	4
4 cups	chicken or vegetable stock	1 L
1/2 cup	whipping cream	120 mL
	salt and freshly ground black pepper to taste	
1/2 cup	sour cream	120 mL

Serves 6 to 8

If you never eat beets, you've got to try this soup. It will convert even the most die-hard beet-hater. It is good either hot or cold; if you're serving it cold, just make it ahead and refrigerate it until serving time.

In a medium saucepan bring salted water to a boil over high heat. Add the beets and cook until tender, approximately 25 minutes. Drain. Chill the beets under cold running water until they're cool enough to handle. Remove the top and bottom and slip off the skins. Cut the beets into quarters.

While the beets are cooking, melt the butter in a large saucepan over medium-high heat. Add the leeks and onion and sauté until they are translucent. Add the potatoes and stock, and bring to a boil. Remove any scum that appears on the surface. Reduce the heat, cover and simmer until the potatoes are tender, approximately 15 minutes.

Remove from the heat, cool a little, add the beets and purée the soup in batches in a blender until very smooth. Return to a clean pot and stir in the cream. Adjust the seasoning with salt and pepper. Reheat the soup to serving temperature.

Ladle into individual warmed serving bowls and garnish each with a dollop of sour cream.

GINGERED CARROT SOUP

Serves 6

The spicy scent of ginger paired with the sweetness of carrots is warming to the heart on a cold winter's night. Carrots are available all year long, but this soup is especially good in winter when fresh produce is looking a little tired, not to mention expensive.

1 Tbsp.	butter	15 mL
1 Tbsp.	vegetable oil	15 mL
2	medium onions, roughly chopped	2
1	leek, white part only, roughly chopped	1
1/4 cup	roughly chopped celery	60 mL
1 Tbsp.	peeled and finely chopped fresh ginger	15 mL
1	large clove garlic, finely chopped	1
3 lbs.	carrots, roughly chopped	1.4 kg
5 cups	chicken or vegetable stock	1.2 L
1/2 cup	whipping cream	120 mL
	salt and freshly ground black pepper to taste	
1/2 cup	sour cream (optional)	120 mL

Heat the butter and oil in a large stockpot over medium-high heat. Add the onion, leek and celery and sauté until the onion is translucent. Do not brown. Add the ginger and garlic and sauté for 1 minute more. Add the carrots and stock and bring to a boil. Remove any scum that comes to the surface. Reduce the heat and simmer until the carrots are tender, approximately 30 minutes. Remove from the heat and allow the soup to cool slightly.

Purée the soup in a blender, a few cups at a time, until it's very smooth. Pour it into a clean pot and add the cream. Bring the soup back to serving temperature over medium heat, stirring often to prevent sticking. Season with salt and pepper. Garnish with sour cream, if desired, and serve immediately.

CAULIFLOWER AND CHEDDAR SOUP

2 cups	coarsely chopped cauliflower	480 mL
2 cups	peeled and diced potato	480 mL
1	large carrot, peeled and diced	1
1	large onion, finely chopped	1
1	large clove garlic, peeled	1
1	bay leaf	1
1 tsp.	fresh thyme	5 mL
4 cups	chicken or vegetable stock	1 L
1 1/2 cups	coarsely chopped cauliflower	360 mL
3/4 cup	milk	180 mL
1 1/2 cups	shredded sharp Cheddar cheese	360 mL
1/2 tsp.	Dijon mustard	2.5 mL
3/4 cup	buttermilk	180 mL
	salt and freshly ground black pepper to taste	
8	fresh thyme sprigs	8

Serves 8

Hearty soups like this one are especially good when the evenings are cool in early autumn.

In a large saucepan combine 2 cups (480 mL) of the cauliflower, the potatoes, carrot, onion, garlic, bay leaf, thyme and stock. Bring to a boil, reduce the heat and simmer for 20 minutes, or until the vegetables are tender. Remove from the heat, discard the bay leaf, and let cool for a few minutes.

In a separate pot, cover the remaining 1 1/2 cups (360 mL) of cauliflower with water and bring to a boil. Simmer for 10 minutes or until the cauliflower is tender. Drain and set aside.

Transfer the mixed vegetables, a few cups at a time, to a food processor or blender and purée it. Return the mixture to a clean pot, place over medium-low heat, add the milk, shredded cheese, Dijon mustard, buttermilk and reserved cauliflower. Heat the soup to serving temperature, stirring frequently. Do not let it boil. Adjust the seasoning with salt and pepper.

Ladle the soup into warm bowls and garnish each bowl with a fresh thyme sprig.

ROASTED CARROT, ONION AND GARLIC SOUP

Serves 8

The roasted vegetables give this soup a lovely, deep, rich flavor. Serve it with crusty country bread for the beginning to a really great meal.

1/4 cup	vegetable oil	60 mL
3 lbs.	carrots, peeled and cut into 1-inch (2.5-cm) pieces	1.4 kg
3	large onions, coarsely chopped	3
	salt and freshly ground black pepper to taste	
1	head garlic	1
2 tsp.	chopped fresh thyme	10 mL
10 cups	chicken or vegetable stock	2.4 L
2 Tbsp.	butter	30 mL
	fresh thyme sprigs (optional)	

Preheat the oven to 350°F (175°C).

Heat all but 1 tsp. (5 mL) of the vegetable oil over medium heat in a large sauté pan. Add the carrots and onions and sauté until the onions are translucent, about 7 to 10 minutes. Season with salt and pepper.

Meanwhile, remove and discard the top 1/4 inch (.6 cm) of the garlic bulb. Place the garlic on a sheet of aluminum foil, pour the reserved 1 tsp. (5 mL) of vegetable oil over it, season with salt and pepper, and twist the foil around the garlic, closing it securely.

Transfer the carrots and onions to a lightly oiled baking sheet. Place the garlic in a corner of the sheet and bake until the vegetables begin to brown, approximately 45 minutes. Stir occasionally.

Transfer the carrots and onions to a saucepan. Squeeze the individual cloves of garlic into the mixture and discard the skins. Add the thyme and stock. Bring to a boil. Remove any scum that comes to the surface. Reduce the heat and simmer until the carrots are soft, approximately 30 minutes. Cool slightly.

In a food processor or blender, process the soup in small batches until it's velvety smooth. Return the soup to a clean pot. Add more stock if the soup is too thick.

Bring the soup to a simmer, stir in the butter, taste and adjust the seasoning. Garnish with thyme sprigs, if desired.

CURRIED BEET SOUP WITH HONEY

1 Tbsp.	vegetable oil	15 mL
1	medium onion, finely chopped	1
1 tsp.	curry powder	5 mL
8	medium beets, peeled and sliced	8
2	medium potatoes, peeled and diced	2
6 cups	chicken stock	1.5 L
1 Tbsp.	honey	15 mL
	salt and freshly ground black pepper to taste	
1/4 cup	sour cream	60 mL
1/4 cup	finely chopped chives	60 mL

Serves 8

An unlikely combination but delicious all the same.

Heat the oil in a large saucepan over medium-high heat. Add the onion and sauté until translucent. Add the curry powder and sauté for 1 minute more. Add the beets, potatoes, stock and honey and bring to a boil. Remove any scum that comes to the surface. Reduce the heat and simmer covered for 30 minutes, or until the vegetables are tender.

Remove from the heat to cool a little. Working in small batches, purée the soup in a blender until smooth. Return the soup to a clean pot and adjust the seasoning with salt and pepper.

Mix the sour cream and chives together. Ladle the soup into serving bowls. Garnish each bowl with a dollop of sour cream and chives.

CREAM OF CARROT AND ORANGE SOUP

Serves 8

Oranges give this soup
a citrus lift, well worth
trying when the winter
doldrums are upon us.

2	medium navel oranges	2
3 Tbsp.	unsalted butter	45 mL
1	small onion, finely chopped	1
6 cups	peeled and thinly sliced carrots	1.5 L
1/2 tsp.	chopped fresh ginger	2.5 mL
1	small garlic clove, finely chopped	1
1 tsp.	ground cumin	5 mL
	few drops hot pepper sauce	
4 cups	chicken stock	1 L
1 cup	half-and-half cream	240 mL
1 Tbsp.	unsalted butter	15 mL
	salt and freshly ground black pepper to taste	
1/2 cup	sour cream (optional)	120 mL

Grate the zest from the oranges and set it aside. You should have about 1 Tbsp. (15 mL). Remove and discard the pith, reserving the fleshy orange sections.

Melt the 3 Tbsp. (45 mL) of butter in a large saucepan over medium-high heat. Add the onion and sauté until it is translucent. Add the carrots and sauté 2 minutes more. Add the ginger, garlic and cumin, and sauté for 30 seconds. Stir in the pepper sauce and chicken stock. Bring to a boil, and add the orange sections and 1/2 Tbsp. (7.5 mL) of the orange zest. Remove any scum that comes to the surface. Reduce the heat and simmer until the carrots are tender, approximately 20 minutes. Remove from the heat and let cool a little before proceeding.

Transfer the soup to a food processor or blender and purée it in small batches. Return the soup to a clean saucepan, add the cream and heat the soup through (do not let it boil). Just before serving, swirl the 1 Tbsp. (15 mL) of butter through the soup.

Taste, and adjust the seasoning with salt and pepper.

Ladle the soup into warm bowls. Garnish each serving with a dollop of sour cream and a sprinkle of the remaining orange zest.

SOME GARNISHES FOR SOUP

- *Add a dollop of sour cream, crème fraîche or yogurt to a puréed or cream soup.*

- *Add a fresh herb sprig—thyme, coriander, parsley, a basil leaf, chive sprigs or lovage leaves, to name a few—to a puréed or cream soup.*

- *Scatter finely chopped fresh herbs around the rim of the soup plate. Coriander and parsley do nicely. Be sure not to put them on top of the soup. Too much may change the flavor of the soup significantly.*

- *Sprinkle a few rose-colored or multicolored cracked peppers over the top of a puréed soup.*

- *A little red pepper coulis or basil pesto can give a lift to a vegetable soup such as minestrone.*

- *Sprinkle the top of the soup with croutons, enhanced with a little garlic, butter and herbs.*

- *Edible flowers, especially nasturtiums, look attractive on the side of a soup plate.*

SMOKY CORN CHOWDER

Serves 6 to 8

Bacon is what gives this down-home chowder its smoky flavor. If you're looking for a completely vegetarian chowder, omit the bacon and substitute 4 tablespoons (60 mL) of butter. You won't have the smoky taste, but it's still a very satisfying chowder.

4	slices bacon, chopped into 1/4-inch (.6-cm) slices	4
1	medium onion, diced	1
1	sweet red pepper, cut into 1/4-inch (.6-cm) dice	1
1	clove garlic, finely chopped	1
3 Tbsp.	flour	45 mL
1 cup	milk	240 mL
1 Tbsp.	vegetable oil	15 mL
2	stalks celery, finely chopped	2
4	potatoes, peeled and cut into 1/2-inch (1.2-cm) cubes	4
7 cups	chicken or vegetable stock	1.75 L
4	ears corn, kernels removed from the cob, or 4 cups (1 L) frozen	4
2 tsp.	Worcestershire sauce	10 mL
1 cup	whipping cream	240 mL
	salt and freshly ground black pepper to taste	
1 Tbsp.	butter	15 mL
1/2 cup	sweet red pepper, seeded and cut into 1/4-inch (.6-cm) dice	120 mL

Sauté the bacon over medium heat in a medium sauté pan for a few minutes; do not brown. Add the onion and sauté until it is translucent. Add the red pepper and sauté for 1 minute. Add the garlic and sauté for 30 seconds. Sprinkle in the flour and stir until the vegetables are evenly covered with flour. Slowly add the milk, stirring to prevent lumps forming. Remove from the heat and set aside.

In a large stockpot heat the oil over medium heat. Add the celery and sauté for 1 minute. Add the potatoes and stock, bring to a boil, and simmer for 10 minutes. Remove any scum that comes to the surface.

Add the corn and Worcestershire sauce and continue to cook for 10 minutes more, or until the potatoes are tender.

Transfer half the potato and corn mixture and half the bacon and onion mixture to a food processor. Add the cream and purée until smooth. Stir the purée into the soup pot and add the rest of the bacon and onion mixture. Heat the soup until it's very hot, but do not boil it. Season with salt and pepper.

Just before serving, stir in the butter. Sprinkle finely chopped red peppers in the center of each serving.

Note: To remove corn kernels from cobs, run the edge of a very sharp knife down the side of the ear away from you. It's very easy and only takes a few seconds for each ear. One medium-size ear of corn yields approximately 1 cup (240 mL) of corn kernels.

TOMATO FIDDLEHEAD SOUP

Serves 4 to 6

Fiddleheads are tightly coiled new fern fronds. When they start to appear at the market it's a sure sign of spring.

2 Tbsp.	olive oil	30 mL
2 Tbsp.	unsalted butter	30 mL
1	large onion, finely chopped	1
1	carrot, peeled and finely diced	1
1	leek, white part only, chopped	1
2	cloves garlic, finely chopped	2
1 tsp.	fresh thyme	5 mL
1 tsp.	chopped fresh basil	5 mL
1	bay leaf	1
1	28-oz. (796-mL) can plum tomatoes, chopped	1
2 cups	chicken or vegetable stock	480 mL
2 cups	fiddleheads, fresh or frozen	480 mL
	salt and freshly ground black pepper to taste	
1/2 cup	plain yogurt or sour cream	120 mL

Heat the oil and melt the butter in a large saucepan over medium-high heat. Add the onion, carrot and leek, and sauté until the onion is translucent; do not brown. Add the garlic, thyme, basil and bay leaf; sauté for 30 seconds, stirring all the time. Add the tomatoes with their juice and the stock, stir well and bring the mixture to a boil. Discard any scum that comes to the surface.

Reduce the heat and simmer for 20 minutes. Add the fiddleheads and simmer for 10 minutes longer, or until the fiddleheads are tender. Adjust the seasoning with salt and pepper. Discard the bay leaf.

Ladle into bowls and garnish each serving with a little yogurt or sour cream.

CREAM OF GARLIC SOUP

3 Tbsp.	butter	45 mL
1	large onion, finely chopped	1
2	heads garlic, cloves separated and roughly chopped	2
2	large potatoes, peeled and thinly sliced	2
3 Tbsp.	all purpose flour	45 mL
3 1/2 cups	chicken stock	840 mL
2	fresh sage leaves	2
2	sprigs fresh thyme	2
	pinch freshly grated nutmeg	
1	bay leaf	1
1 cup	whipping cream	240 mL
	salt and freshly ground black pepper to taste	
8	fresh sage leaves (optional)	8

Serves 4

This soup delivers a burst of garlic flavor in every spoonful.

Melt the butter in a large saucepan over medium heat. Add the onion and sauté for 2 minutes. Add the garlic and sauté over low heat for 10 minutes. Do not let it brown. Add the potatoes, sauté for 1 minute, then add the flour and sauté for 2 minutes more.

Gradually whisk in the stock. Add the 2 sage leaves, thyme, nutmeg and bay leaf. Bring to a boil. Remove any scum that comes to the surface, reduce the heat and simmer for 15 minutes, or until the potatoes are tender.

Remove from the heat. Discard the sage leaves, thyme and bay leaf. Transfer to a food processor or blender and purée the soup in batches until it's smooth. Pass the soup through a fine sieve. Return the soup to a clean pot. Add the cream and reheat the soup slowly; do not let it boil. Adjust the seasoning with salt and pepper.

Ladle into warm bowls and garnish each with a fresh sage leaf, if desired.

CLASSIC LEEK AND POTATO SOUP

Serves 6

Still considered a little exotic in our cuisine, leeks are the mainstay of French cooking. This rich and satisfying soup is as smooth as silk. Serve it warm in the winter and in the summer serve it chilled—when its name changes to the elegant Vichyssoise.

2 Tbsp.	butter	30 mL
3	medium leeks, white part only, washed and finely chopped	3
1	medium onion, finely chopped	1
4	medium potatoes, peeled and thinly sliced	4
4 cups	chicken stock	1 L
1 tsp.	fresh thyme	5 mL
1/2 tsp.	fresh parsley	2.5 mL
1/4 tsp.	ground white pepper	1.2 mL
1 cup	whipping cream	240 mL
1 Tbsp.	butter	15 mL
	salt to taste	
1/4 cup	finely chopped chives (optional)	60 mL

Melt the butter in a large saucepan over medium-high heat. Add the leek and onion and sauté lightly until they are translucent. Add the potatoes, stock, thyme, parsley and pepper. Bring to a boil. Remove any scum that comes to the surface. Cover, reduce the heat and simmer until tender, approximately 15 minutes.

Remove from the heat, cool a little, then process in a blender until very smooth. Transfer to a clean pot and return to the heat. Add the cream and butter, stirring until the butter is melted. Adjust the seasoning with salt and white pepper.

Ladle the soup into warm individual bowls and sprinkle a few chopped chives over each, if desired.

If you're serving the soup cold, refrigerate it for at least 4 hours or preferably overnight.

MINESTRONE VERDI

2 Tbsp.	olive oil	30 mL
1	large onion, finely chopped	1
1/2 cup	diced celery	120 mL
1/4 cup	diced carrot	60 mL
1/4 cup	diced sweet green pepper	60 mL
1 cup	red potatoes, scrubbed and diced	240 mL
1	clove garlic, finely chopped	1
1/4 cup	dry white wine	60 mL
8 cups	chicken stock	2 L
	pinch saffron	
	sprig each fresh thyme and parsley	
1 cup	fresh green beans, trimmed and cut into 1-inch (2.5-cm) lengths	240 mL
1/2 cup	fresh or frozen lima beans	120 mL
3 Tbsp.	orzo	45 mL
1 cup	escarole, tender inside leaves, packed	240 mL
1 cup	diced zucchini	240 mL
1/2 cup	asparagus tips	120 mL
1/2 cup	fresh or frozen peas	120 mL
	salt and freshly ground black pepper to taste	
1/2 cup	grated Parmesan cheese	120 mL

Serves 6

Not the classic minestrone with tomatoes, this soup has the hidden surprise of saffron.

Heat the oil over medium-high heat in a large heavy stockpot. Add the onion, celery and carrot. Sauté until the onion is translucent. Add the green pepper and sauté 1 minute more. Add the potato and garlic; sauté for 30 seconds. Add the white wine and cook until the liquid is reduced by half.

Add the chicken stock, saffron, thyme and parsley and bring to a boil. Reduce the heat to medium and add the green beans, lima beans and orzo. Cook for 10 minutes, or until the beans are tender. Add the escarole, zucchini, asparagus tips and peas. Simmer for 5 minutes. Season with salt and pepper.

Ladle into serving bowls. Garnish with freshly grated Parmesan cheese.

CLASSIC FRENCH ONION SOUP

Serves 8

Slow-cooking the onions and shallots is the secret to this timeless soup. Change the beef stock to chicken stock and you have Swiss onion soup.

1/4 cup	unsalted butter	60 mL
6 cups	thinly sliced onions	1.5 L
3	shallots, thinly sliced	3
2	cloves garlic, finely chopped	2
1/2 tsp.	salt	2.5 mL
	freshly ground black pepper to taste	
	pinch sugar	
1 tsp.	dried thyme	5 mL
1	bay leaf	1
2 Tbsp.	cognac	30 mL
3/4 cup	dry white wine	180 mL
1/3 cup	flour	80 mL
8 cups	beef stock	2 L
1	baguette, cut into 1-inch (2.5-cm) slices	1
1 cup	grated Gruyère cheese	240 mL
1 cup	grated Parmesan cheese	240 mL

Melt the butter in a large saucepan over medium heat. Add the onions, shallots and garlic and cook for 15 minutes, stirring frequently. Add the salt, pepper, sugar, thyme and bay leaf. Continue to cook for 45 minutes over low heat, stirring frequently, until the onions are a deep golden color. Increase the heat to medium-low. Add the cognac and continue to cook until it's evaporated, about 1 minute. Add the white wine and cook until it's reduced by half, about 5 minutes.

Sprinkle the flour over the onion mixture and stir for 2 minutes. Add the beef stock. Simmer the soup for 45 minutes. Discard the bay leaf. Taste, and adjust the seasoning.

Place the baguette slices under the broiler and toast both sides. Mix the cheeses together.

Ladle the soup into ovenproof bowls. Float a baguette slice in each bowl on top of the soup, and cover generously with the cheese mixture. Place the soup under the broiler until the cheese is bubbly and brown on top. Serve immediately.

CREAM OF PARSNIP SOUP WITH SHERRY

1/4 cup	unsalted butter	60 mL
6	medium parsnips, peeled and roughly chopped	6
3	large shallots, finely chopped	3
3	leeks, white part only, washed and roughly chopped	3
1	rib celery, roughly chopped	1
1 tsp.	chopped fresh thyme	5 mL
1	bay leaf	1
3 Tbsp.	dry sherry	45 mL
5 cups	chicken or vegetable stock	1.25 L
1 Tbsp.	dry sherry	15 mL
1/4 cup	whipping cream	60 mL
	pinch freshly grated nutmeg	
	salt and freshly ground black pepper to taste	

Serves 5 to 6

This would be a great starter to a turkey dinner —the hint of sherry gives it a festive touch. To garnish it, thin a little Basil Pesto (page 216) with olive oil and drizzle it over the soup.

Melt the butter in a large saucepan over medium-high heat. Add the parsnips, shallots, leeks and celery, and sauté until the vegetables are soft, approximately 5 minutes. Add the thyme, bay leaf and the 3 Tbsp. (45 mL) sherry. Sauté until the liquid has evaporated. Add the stock and bring to a boil. Remove any scum that comes to the surface. Reduce the heat and simmer for 20 minutes, or until the vegetables are very soft. Remove from the heat and discard the bay leaf.

When the soup has cooled slightly, purée it in batches in a blender or food processor until it is very smooth. Return the soup to a clean pot and add the remaining tablespoon (15 mL) of sherry and the whipping cream. Heat through and season with a little nutmeg, salt and pepper.

Brandied Cream of Mushroom Soup

Serves 6

Older mushrooms that are beginning to darken give this soup a richer mushroom flavor. For something more exotic, use half wild mushrooms and half domestic mushrooms.

1/4 cup	unsalted butter	60 mL
1/4 cup	finely chopped onion	60 mL
1/4 cup	finely chopped celery	60 mL
2	shallots, finely chopped	2
4 cups	mushrooms, thinly sliced	1 L
2 Tbsp.	brandy	30 mL
4 cups	chicken stock	1 L
1/4 cup	unsalted butter	60 mL
1/4 cup	flour	60 mL
2 cups	cold milk	480 mL
1	bay leaf	1
	grating of fresh nutmeg	
1/2 cup	whipping cream	120 mL
	salt and freshly ground black pepper to taste	
2 Tbsp.	unsalted butter	30 mL
2 cups	mushrooms, thinly sliced	480 mL
1 Tbsp.	brandy	15 mL

Melt 1/4 cup (60 mL) of butter in a medium saucepan over medium-high heat. Add the onion and celery; sauté until they begin to soften, approximately 3 minutes. Add the shallots and sauté for 1 minute. Add the 4 cups (1 L) of mushrooms and sauté for 3 minutes. Add the 2 Tbsp. (30 mL) of brandy and continue to cook until all the liquid has evaporated. Stir in the chicken stock and continue to simmer for 20 minutes. Let cool slightly. Transfer the mixture to a food processor or blender and purée in small batches. Set aside.

Melt the remaining 1/4 cup (60 mL) of butter in a separate saucepan over medium heat. When the butter is bubbly, add the flour and stir until it's well combined and lump-free. Gradually pour in the cold milk, stirring and cooking until the sauce is smooth. Add the bay leaf and nutmeg. Continue to cook over low heat until thick. If the sauce is lumpy, pour it through a fine mesh sieve and push the lumps through with the back of a wooden spoon.

Slowly add the puréed mushroom mixture to the white sauce, stirring until it's well combined. Continue to cook over low heat. Add the cream, and adjust the seasoning with salt and pepper. Remove the bay leaf.

Melt the 2 Tbsp. (30 mL) of butter in a sauté pan over medium-high heat. Quickly sauté the remaining 2 cups (480 mL) of mushrooms. Season with salt and pepper. When the mushrooms are wilted, add the remaining 1 Tbsp. (15 mL) of brandy and continue to cook until all the liquid has evaporated.

Ladle the soup into warm bowls. Place a few sautéed mushrooms on top of each serving. Serve immediately.

SPLIT PEA SOUP

Serves 8

The French call it Potage Saint-Germain and it's been a mainstay in French cuisine for generations.

2 cups	dried split peas	480 mL
6 cups	water	1.5 L
1 tsp.	salt	5 mL
8	slices bacon, finely chopped	8
1 Tbsp.	butter	15 mL
1	onion, finely chopped	1
1	carrot, peeled and finely chopped	1
2	leeks, white parts only, cleaned and finely chopped	2
1 cup	shredded Boston lettuce	240 mL
1	bay leaf	1
1/4 tsp.	dried thyme	1.2 mL
1 cup	fresh or frozen peas	240 mL
2 cups	water	480 mL
1 1/2 cups	chicken stock	360 mL
1/2 cup	half-and-half cream	120 mL
1 tsp.	sugar	5 mL
	salt and black pepper to taste	
2 Tbsp.	butter	30 mL

Place the split peas in a large saucepan, add 6 cups (1.5 L) of the water and the salt. Bring to a boil, reduce the heat and simmer covered for 45 minutes, stirring occasionally, until most of the water is absorbed.

Place the bacon and the 1 Tbsp. (15 mL) of butter in a large skillet over medium-high heat. Sauté the bacon until it starts to crisp. Add the onion, carrot and leeks and sauté until they start to soften. Add the lettuce, bay leaf and thyme, and sauté for 1 minute. Add the sautéed vegetables to the split peas along with the fresh or frozen peas. Add the remaining 2 cups (480 mL) of water, cover and continue to simmer 1 hour longer.

Discard the bay leaf. Transfer the soup to a food mill and purée. Transfer the soup to a clean pot and return it to the heat. Add the stock, cream and sugar and reheat. Season with salt and pepper. Just before serving, stir in the remaining 2 Tbsp. (30 mL) of butter.

BASIL AND ROASTED RED PEPPER SOUP

4 Tbsp.	butter	60 mL
4	large leeks, white parts only, washed and finely chopped	4
3	large onions, finely chopped	3
2	cloves garlic, finely chopped	2
6 cups	chicken stock	1.5 L
6	large potatoes, peeled and cut into 1/4-inch (.6-cm) dice	6
2 tsp.	chopped fresh thyme	10 mL
1 tsp.	chopped fresh parsley	5 mL
1 cup	whipping cream	240 mL
	salt and freshly ground black pepper to taste	
1/2 cup	Basil Pesto (page 216)	120 mL
3	sweet red peppers, roasted (see page 5)	3
10	basil leaves for garnish	10

Serves 10

I like to call this soup the dynamic duo. It is almost as beautiful as it is delicious. You begin by making one large pot of soup and then dividing the pot in half. To one half add Basil Pesto and to the other half puréed roasted red peppers. To serve, pour the two soups into one bowl. This soup is a little time-consuming, but it can be made a day ahead and reheated just before serving.

Melt the butter in a large saucepan over medium-high heat. Add the leeks and onions and sauté until the onion is translucent. Add the garlic and sauté for 30 seconds. Add the chicken stock, potatoes, thyme and parsley and bring to a boil. Remove any scum that comes to the surface of the soup. Reduce the heat and simmer for 30 minutes. Remove from the heat and cool slightly.

Working in small batches, process the soup in a food processor or blender until it's very smooth. Return it to the pot, add the cream and adjust the seasoning with salt and pepper.

Divide the soup into two containers. To the first add the Basil Pesto. Stir and keep warm.

Purée the roasted red peppers in a food processor or blender. Add the purée to the second container of soup.

The key to serving the soup is pouring into opposite sides of the bowl at the same time. Pour the soup into measuring cups with spouts to make the job easier. Garnish with a fresh basil leaf.

CREAM OF SWEET RED PEPPER SOUP

Serves 4 to 6

This is a very quick and easy soup to prepare. To make it even faster, you can roast the peppers a day ahead and store them in the refrigerator until you're ready to use them.

1 Tbsp.	vegetable oil	15 mL
3	shallots, finely chopped	3
2	cloves garlic, finely chopped	2
1 Tbsp.	chopped fresh thyme	15 mL
1	bay leaf	1
3 cups	chicken or vegetable stock	720 mL
10	medium sweet red peppers, roasted (see page 5) and cut into slices	10
1/2 cup	whipping cream	120 mL
2 tsp.	red wine vinegar	10 mL
1/8 tsp.	cayenne pepper	.5 mL
	salt and freshly ground black pepper to taste	
6	basil leaves (optional)	6

Heat the oil in a saucepan over medium heat. Sauté the shallots until translucent, add the garlic, thyme and bay leaf and sauté for 30 seconds. Add the stock and all but 4 or 6 slices of roasted red pepper. Simmer uncovered until the peppers are very soft, approximately 20 minutes. Remove any scum that comes to the surface.

Remove the bay leaf. Working in small batches, process the mixture in a blender or food processor until it's very smooth. Return the purée to a clean pot and add the cream, vinegar and cayenne pepper. Reheat the soup. If it is too thick, thin it with additional stock. Season with salt and pepper.

Garnish each serving with a thin strip of reserved red pepper and a basil leaf, if desired.

MOROCCAN PUMPKIN SOUP

1 Tbsp.	olive oil	15 mL
1 Tbsp.	butter	15 mL
1	medium onion, finely chopped	1
1	clove garlic, finely chopped	1
1 1/2 Tbsp.	mild curry powder	22.5 mL
1 Tbsp.	cumin	15 mL
1/4 tsp.	cayenne pepper	1.2 mL
1 tsp.	powdered ginger	5 mL
3 cups	chicken stock	720 mL
2 cups	coconut milk	480 mL
2 cups	canned pumpkin purée	480 mL
1	bay leaf	1
1/2 tsp.	grated lemon zest	2.5 mL
1/2 tsp.	grated orange zest	2.5 mL
	salt and freshly ground black pepper to taste	
1/2 cup	plain yogurt (optional)	120 mL
1 Tbsp.	grated orange zest (optional)	15 mL
1 Tbsp.	grated lemon zest (optional)	15 mL

Serves 6

Not just for pies any more, pumpkin has arrived as a vegetable with numerous possibilities. Try this warm and fragrant soup when the nip of fall is in the air.

Heat the olive oil and butter in a large saucepan over medium heat. Add the onion and sauté until translucent. Add the garlic and sauté 30 seconds more. Add the curry powder, cumin, cayenne pepper and ginger. Sauté for 1 minute. Add the stock, coconut milk, pumpkin purée, bay leaf and the 1/2 tsp. (2.5 mL) lemon and orange zest.

Bring the soup to a boil. Discard any scum that comes to the surface. Reduce the heat and simmer uncovered for 30 minutes. Remove from the heat. Discard the bay leaf. If a smoother soup is required, purée in a food processor or electric blender. Taste and adjust the seasoning with salt and pepper.

Pour the soup into warm bowls. If desired, garnish with a swirl of plain yogurt and sprinkle with lemon and orange zest.

Potato, Cheddar and Beer Soup

*Serves 6 to 8
as a main course*

This is a hearty, stick-to-your-ribs kind of soup—English pub-style fare that's ideal on a cold winter evening. For those who like their soup spicy, pass around a bottle of hot pepper sauce.

5	large potatoes, peeled and thinly sliced	5
2 1/2 cups	beer or ale	600 mL
4 cups	chicken or vegetable stock	1 L
1	bay leaf	1
1/2 tsp.	dried basil	2.5 mL
1/2 tsp.	dried thyme	2.5 mL
4 Tbsp.	unsalted butter	60 mL
1	large onion, diced	1
1	stalk celery, diced	1
1	large carrot, peeled and diced	1
1	leek, white part only, finely chopped	1
1/2 tsp.	salt	2.5 mL
4	cloves garlic, finely chopped	4
1	14-oz. (398-mL) can tomatoes with juice, chopped	1
2 tsp.	Worcestershire sauce	10 mL
2 Tbsp.	flour	30 mL
1 cup	whipping cream	240 mL
12 oz.	shredded sharp Cheddar cheese	340 g
	salt and freshly ground black pepper to taste	

Place the potatoes, beer or ale, stock, bay leaf, basil and thyme in a large pot over medium-high heat. Bring to a boil. Remove any scum that appears on the surface. Cover and simmer for approximately 20 minutes, or until the potatoes are tender.

Meanwhile, melt the butter in a large frying pan over medium-high heat. Add the onion, celery, carrot and leek and sprinkle with the salt. Sauté until the onions are translucent. Add the garlic and sauté for 30 seconds longer. Stir in the tomatoes and Worcestershire sauce and cook until all the liquid evaporates, approximately 20 minutes. Sprinkle the mixture with the flour and stir until it's well combined. Gradually add the cream. Cook and stir until the mixture is smooth and thick.

Add the vegetable-cream mixture to the potatoes and stock. Keeping the heat very low, stir in the cheese. Heat the soup until the cheese has melted, about 20 minutes. Stir often, as the soup is very thick and tends to stick to the bottom of the pot. Discard the bay leaf. Taste and adjust the seasoning with more salt, if needed, and pepper.

MEDLEY OF THREE SQUASH SOUP

Serves 6 to 8

Don't tell anyone who doesn't like squash what's in this soup— they'll never guess. Try any combination of squashes; if you have a bumper crop of any one variety, that works nicely as well. This soup stores well in the refrigerator, so it can be made a day ahead.

1/2 lb.	each butternut, acorn and Hubbard or turban squash	225 g
1 Tbsp.	olive oil	15 mL
1	small onion, finely chopped	1
1 lb.	carrots, peeled and thinly sliced	455 g
1 tsp.	each paprika and turmeric	5 mL
2 tsp.	each ground cumin and coriander	10 mL
1	clove garlic, finely minced	1
4 cups	chicken stock	1 L
	salt and freshly ground black pepper to taste	
1/2 cup	plain yogurt or sour cream (optional)	120 mL

Preheat the oven to 350°F (175°C). Cut the squash into large pieces and discard the seeds. Bake until tender, about 45 minutes to 1 hour.

Heat the olive oil in a large saucepan over medium heat. Add the onion and carrot and sauté until the onion is translucent. Stir in the paprika, turmeric, cumin, coriander and garlic, and cook 1 minute longer. Add the stock. Scoop the baked squash from the shell and stir it into the soup.

Bring the soup to a boil. Remove any scum that appears on the surface. Cover and gently simmer until the carrots are tender, approximately 30 minutes. Remove from the heat and cool slightly. Purée the soup in a blender or food processor until it's very smooth. Strain into a clean saucepan and reheat the soup to serving temperature. Season with salt and pepper and thin with a little more stock if the soup is too thick.

Ladle the soup into warm serving bowls and garnish with a dollop of plain yogurt or sour cream, if desired.

CREAM OF TOMATO SOUP

3 Tbsp.	butter	45 mL
1	large onion, finely chopped	1
1	stalk celery, finely chopped	1
1	carrot, peeled and finely chopped	1
1	large clove garlic, finely chopped	1
1 Tbsp.	tomato paste	15 mL
2	whole cloves	2
1	bay leaf	1
1 1/2 tsp.	dried basil	7.5 mL
1 tsp.	dried thyme	5 mL
4 lbs.	fresh tomatoes (approximately 16 medium), cored and roughly chopped	1.8 kg
2 tsp.	brown sugar	10 mL
1 tsp.	salt	5 mL
1/2 cup	whipping cream salt and freshly ground black pepper to taste	120 mL

Serves 6 to 8

Save making this soup for the end of summer when tomatoes are plentiful and their flavor superb. A garnish of homemade croutons finishes it off perfectly.

Melt the butter in a large saucepan with a tight-fitting lid over medium-high heat. Add the onion, celery and carrot and sauté for 3 minutes. Cover and sauté 5 minutes more. Add the garlic and sauté for 30 seconds. Add the tomato paste and sauté until the paste turns a deep rust color. Add the cloves, bay leaf, basil, thyme and fresh tomatoes.

Bring the mixture to a boil, add the brown sugar and salt, reduce the heat and simmer until the vegetables are soft, approximately 40 minutes. Remove from the heat.

Discard the cloves and bay leaf. Purée the soup in a food mill. Discard the seeds and skins. Transfer the soup to a clean pot. Reheat the soup and add the cream. Adjust the seasoning with salt, if necessary, and pepper.

CREAM OF ZUCCHINI SOUP WITH RED PEPPER SAUTÉ

Serves 4 to 6

What can you do with all those zucchinis coming up in the garden? Grate them and freeze them in recipe-sized batches— they'll make great soup later on.

3 Tbsp.	vegetable oil	45 mL
1	medium onion, finely chopped	1
1	sweet green pepper, finely chopped	1
2 1/2 cups	finely grated zucchini	600 mL
1 cup	chicken or vegetable stock	240 mL
1/4 cup	butter	60 mL
1/4 cup	all purpose flour	60 mL
2 1/2 cups	milk	600 mL
1 tsp.	Worcestershire sauce	5 mL
1/2 tsp.	dried dill	2.5 mL
1 cup	shredded sharp Cheddar cheese	240 mL
	salt and freshly ground black pepper to taste	
2 tsp.	butter	10 mL
1	small clove garlic, finely chopped	1
1	sweet red pepper, finely chopped	1

Heat the oil in a medium saucepan over medium-high heat. Add the onion and sauté for 2 minutes. Add the green pepper and sauté until tender, about 3 minutes. Add the zucchini and sauté for 1 minute. Add the stock and bring to a boil. Remove any scum that appears on the surface. Reduce the heat and simmer for 5 minutes.

In a separate saucepan melt the 1/4 cup (60 mL) butter over medium heat. Add the flour and stir, making a roux. Slowly add the milk, stirring constantly so no lumps form. Continue stirring until all the milk is incorporated and the sauce is thick. Pass the sauce through a fine sieve, using a wooden spoon to push any lumps through. Return it to a clean pot and add the Worcestershire sauce, dill and Cheddar cheese. Continue to cook over low heat, stirring frequently until the cheese has melted.

Add the vegetable mixture and continue to cook over low heat. Season with salt and pepper. Keep the soup warm, but do not let it boil, while making the garnish.

In a small sauté pan melt the 2 tsp. (10 mL) of butter over medium-high heat. Quickly sauté the garlic for 30 seconds. Add the red pepper and sauté until tender, approximately 3 minutes. Season with salt and pepper.

Ladle the soup into bowls, and garnish each bowl with the sautéed red pepper.

ROASTED TOMATO AND RED PEPPER SOUP

Serves 6

The best time to make this soup is in the summer, when tomatoes are at their most flavorful and sweet bell peppers are plentiful. Roasting the vegetables gives the soup its unique flavor.

3 1/2 lbs.	ripe tomatoes, cored and quartered	1.6 kg
3 lbs.	sweet red peppers, quartered	1.4 kg
1	large Spanish onion, peeled and cut into 8 pieces	1
6	carrots, peeled and quartered lengthwise	6
8	cloves garlic, peeled	8
3 Tbsp.	olive oil	45 mL
2	sprigs each fresh thyme, oregano and parsley	2
	salt and freshly ground black pepper to taste	
6 cups	chicken or vegetable stock	1.5 L
6	fresh basil leaves, julienned	6
6	fresh whole basil leaves	6

Preheat the oven to 400°F (200°C). Place the tomatoes, red peppers, onion, carrots and garlic in a large bowl. Toss with the olive oil and add the herb sprigs. Season with salt and pepper and toss the vegetables again. Transfer to a large baking dish or sheet. Roast the vegetables in the oven for 40 to 50 minutes, turning frequently to make sure they are browning evenly.

When the vegetables are tender, remove them from the oven. Remove any charred skin from the red pepper and discard. Dice the carrot and set it aside.

When the vegetables have cooled a little, place them in the food processor (excluding the carrot). Add a little stock and process until the vegetables are puréed.

Place the puréed vegetables, the remainder of the stock and the diced carrots in a saucepan. Add the julienned basil. Heat the soup to serving temperature and serve immediately. Garnish with whole basil leaves.

SALADS

*The list for salad ingredients is endless.
Traditionally in many households, a salad meant a
cold offering of lettuce, tomato wedges, cucumber slices
and the possible addition of a few chopped green onions,
all tossed together and served with gooey bottled dressings.
These salads were only to be served on hot summer days.*

*In today's cuisine, salad can be the focus of a meal—
served as an entrée, hot or cold. The ingredients
may include grilled meats and vegetables, nuts,
beans, cheese, herbs and flowers, to name a few.
But best of all, salads now have no seasonal bounds.*

Asparagus, Snow Pea and Chèvre Salad with Balsamic Vinaigrette

Serves 8 to 10

This composed salad is destined to become a classic. It takes a little time to arrange but it's well worth the effort. It's not too often you get to try your design artistry with a salad. The purple radicchio and red leaf lettuce are lovely accents to this dish.

1/2 lb.	asparagus spears	225 g
1 cup	snow peas	240 mL
1	large red leaf lettuce	1
1	radicchio	1
1	endive	1
2	ripe avocados	2
3 oz.	chèvre	85 g
1/4 cup	pine nuts, toasted (see page 183)	60 mL
1 recipe	Balsamic Vinaigrette	1 recipe

Blanch the asparagus spears in salted boiling water for 3 to 4 minutes, just until they're beginning to become tender. Drain and refresh under cold water. Drain and set aside.

Blanch the snow peas, using the same water, for 1 minute. Drain and refresh under cold water. Drain and set aside.

To make one large salad, line an oval platter with the largest of the lettuce leaves. Tear the remainder of the leaves and the radicchio. Toss them together and place in the center of the platter. Next, slice the bottom off the endive and arrange the leaves around the perimeter of the platter. Remove the skin from the avocados, slice them into thin wedges and place them around the perimeter. Cut the asparagus spears in half crosswise and arrange the tips like spokes on a wheel around the platter. Add the bottom halves of the asparagus to the lettuce and radicchio in the center. Arrange the snow peas around the perimeter. Crumble the goat cheese over the salad and scatter the toasted pine nuts on top. (Alternately, you may want to compose the salad on individual serving plates.)

Just before serving, spoon the vinaigrette over the salad, or pass it around for your guests to help themselves.

Balsamic Vinaigrette

1/4 cup	balsamic vinegar	60 mL
	salt and freshly ground black pepper	
	to taste	
	pinch brown sugar	
1 tsp.	Dijon mustard	5 mL
1	clove garlic, finely chopped	1
1/2 tsp.	finely chopped fresh basil	2.5 mL
3/4 cup	extra virgin olive oil	180 mL

In a small bowl or jar with a tight-fitting lid, combine the vinegar, salt, pepper and sugar and stir until the salt and sugar are dissolved. Add the mustard, garlic, basil and olive oil. Stir well.

FOUR-BEAN SALAD

Serves 6 to 8

The unusual blend of spices in the vinaigrette gives a new twist to an old standard. Cumin is the star, lending this dish a real southwestern flavor.

1	19-oz. (540-mL) can navy beans	1
1	19-oz. (540-mL) can kidney beans	1
1	19-oz. (540-mL) can chick peas	1
1	19-oz. (540-mL) can black beans	1
1	sweet green pepper, cut into 1/4-inch (.6-cm) dice	1
1	sweet red pepper, cut into 1/4-inch (.6-cm) dice	1
3	green onions, chopped into 1/4-inch (.6-cm) pieces	3
1/3 cup	white wine vinegar	80 mL
2 tsp.	lime juice	10 mL
	salt and freshly ground black pepper to taste	
1	clove garlic, finely chopped	1
1 Tbsp.	sugar	15 mL
2 tsp.	ground cumin	10 mL
1 tsp.	Dijon mustard	5 mL
1/4 tsp.	red pepper flakes	1.2 mL
2/3 cup	vegetable oil	160 mL
1/4 cup	finely chopped chives (optional)	60 mL

Drain the cans of beans and rinse the beans under cold running water. Drain and place in a serving bowl. Add the red and green peppers and green onions.

In a small bowl whisk together the vinegar, lime juice, salt and pepper, garlic, sugar and cumin. Whisk in the Dijon mustard, red pepper flakes and vegetable oil. Pour the dressing over the bean mixture. Toss all the ingredients well. Garnish with chopped chives, if desired.

CARROT AND ORANGE SALAD WITH BLACK MUSTARD SEED

1	navel orange	1
3 cups	peeled and grated carrots	720 mL
2 Tbsp.	lemon juice	30 mL
1/2 tsp.	sugar	2.5 mL
1/2 tsp.	salt	2.5 mL
2 Tbsp.	vegetable oil	30 mL
1 tsp.	black mustard seed	5 mL

Serves 6

Grate the zest from the orange, reserving 1 Tbsp. (15 mL). With a sharp knife remove the remaining skin and white pith. Slice the orange crosswise into 1/8-inch (.3-cm) slices. Stack the slices and cut them into 6 wedges.

Place the oranges, carrots, lemon juice, reserved orange zest, sugar and salt in a large bowl. Mix well.

Heat the vegetable oil in a medium skillet over medium-high heat. Add the mustard seeds. When they begin to pop, remove from the heat and add them to the carrot mixture. Toss well and refrigerate until serving time.

A lively, fresh-tasting salad, this goes well with Indian curries or can take the place of a traditional coleslaw. This salad can be made a day ahead and stored in an airtight container in the refrigerator.

Roasted Beet Salad with Fresh Pears and Walnuts over Baby Greens

Serves 4

Roasting beets gives them an extra sweetness and depth of flavor that can't be matched when boiling them. The color alone makes this dish especially appetizing. Beets have been a part of classic French cuisine for generations; this recipe is a new spin on a tried-and-true theme. For a really rich salad, crumble a little Roquefort cheese over top.

5	medium beets, 2 inches (5 cm) in diameter	5
2 Tbsp.	red or white wine vinegar	30 mL
	salt and freshly ground black pepper to taste	
2 tsp.	Dijon mustard	10 mL
1 tsp.	honey	5 mL
1/2 cup	extra virgin olive oil	120 mL
2	ripe pears,	2
1/2	lemon, juice only	1/2
4 cups	loosely packed baby salad greens	1 L
2	endives	2
1/2 cup	walnut halves, toasted (see page 183)	120 mL

Preheat the oven to 375°F (190°C). Line a metal pie plate with 1/2 inch (1.2 cm) of coarse salt. The salt provides a bed for the beets to rest on and prevents them from scorching on the bottom.

Remove the greens from the top of the beets. Arrange the beets on top of the salt, cut side down to ensure they cook evenly and cannot roll around. Bake until tender, approximately 60 to 80 minutes, depending on the size.

Cool the beets on a rack. When they are cool enough to handle, slip the skins off. Dice or julienne the beets and set aside.

To make the vinaigrette, whisk the vinegar, salt and pepper in a bowl until the salt has dissolved. Add the mustard and honey and continue to whisk until well blended. Add the oil in a slow, steady stream until the vinaigrette thickens slightly.

Add the beets to half the vinaigrette to marinate for at least 30 minutes. Reserve the other half of the vinaigrette for serving.

Slice the pears in half and remove the cores with a small spoon or melon baller. Cut them into thin slices from top to bottom. Squeeze the lemon juice over the slices to prevent discoloration.

Place a bed of baby greens on each plate. Slice the bottom off the endives and place the leaves on top of the greens. Arrange the sliced pears around the center of the greens, making a nest for the beets. Place the beets in the center. Sprinkle with walnuts. Serve with the reserved vinaigrette.

TIPS FOR SALADS

- *Dissolve the sugar and salt in the vinegar or lemon juice before adding the oil. Once the oil is added to the mixture they do not dissolve as easily.*

- *Mustard is an emulsifier; when added to salad dressings it makes them thick.*

- *Use just enough dressing to moisten the salad. Pass extra dressing for guests to help themselves.*

- *Composed salads take only a few more minutes to bring together and yet they are so elegant, making a feast for the eyes as well as the palate.*

- *Think of texture as well as taste. Nuts, sprouts, beans and seeds make a salad even more interesting.*

- *Color is another element: flower petals, grilled vegetables, shredded carrots and diced cooked beets all brighten the bowl.*

- *For formal dinner gatherings, arrange salads on chilled individual salad plates.*

- *Store unused portions of salad dressing in the refrigerator. If garlic is present in the mixture, use it in a few days. Garlic goes rancid quickly when added to oil. Olive oil solidifies when refrigerated, but it quickly returns to a liquid consistency when brought back to room temperature.*

GRILLED BREAD SALAD WITH ANCHOVY VINAIGRETTE

Serves 6

A Mediterranean main course salad for those sunny summer days.

6	thick slices rustic country-style bread, such as sourdough or levain	6
2	cloves garlic, halved	2
1	sweet red pepper, cut into 6 pieces	1
1	sweet yellow pepper, cut into 6 pieces	1
1/4 cup	lemon juice	60 mL
6	anchovy fillets, soaked in water for 10 minutes, drained, patted dry and mashed	6
1	shallot, finely chopped	1
1/2 cup	extra virgin olive oil salt and freshly ground black pepper to taste	120 mL
1/2 lb.	goat cheese, semi-soft	225 g
6 cups	mesclun (mixed baby greens)	1.5 L
6	small radishes, tops on, halved	6
1/2 cup	black or green olives	120 mL
1	12-oz. (340-mL) jar marinated artichoke hearts, drained	1
1	lemon, cut into 6 wedges	1

Cut each slice of bread in half on the diagonal. Grill the bread until lightly browned on each side. Rub both sides with the cut side of the garlic halves and set aside.

Heat the barbecue to medium-high.

Lightly brush the red and yellow pepper slices with oil. Grill until marked and just beginning to soften. Remove from the grill and set aside.

In a small bowl whisk together the lemon juice, anchovies, shallot and oil. Season with salt and pepper. Set aside.

Crumble the goat cheese over the prepared bread slices. Place under the broiler until the cheese begins to melt.

Place a handful of the greens onto each salad plate. Place two pieces of grilled bread on top of the lettuce. Arrange the radishes, olives, artichoke hearts, grilled peppers and a lemon wedge on each plate. Drizzle a spoonful of the vinaigrette over the bread. Pass the remaining sauce for guests to help themselves.

BROCCOLI SALAD WITH SUNFLOWER SEEDS

4 cups	broccoli heads and stems, chopped into 1-inch (2.5-cm) pieces	1 L
1/4 cup	finely chopped red onion	60 mL
1/4 cup	chopped celery	60 mL
1/4 cup	raisins	60 mL
1/2 cup	raw unsalted sunflower seeds	120 mL
1/2 lb.	bacon, cooked and crumbled	225 g
1/4 cup	white sugar	60 mL
3/4 cup	prepared salad dressing or Homemade Mayonnaise (page 214)	180 mL
2 tsp.	cider vinegar	10 mL
	salt and freshly ground black pepper to taste	

Serves 6 to 8

Why the fuss over broccoli? Most of us long ago learned to love it, although not the way our mothers cooked it—boiled to within an inch of its life. In a salad, broccoli has a wonderfully crunchy texture and a fresh green taste.

Place the chopped broccoli in a large serving bowl. Mix in the onion, celery, raisins, sunflower seeds and bacon.

Combine the sugar with the salad dressing or mayonnaise. (Use a low-fat dressing if desired.) Add the vinegar and season with salt and pepper. Pour over the salad and toss until well combined. Refrigerate for a few hours before serving. This salad is even better the next day and stores well for up to 3 days.

CAESAR SALAD WITH GARLIC HERB CROUTONS

Serves 8

Garlic lovers will make this creamy Caesar salad dressing over and over again. Don't be afraid of the anchovies; they round out the flavor of an authentic Caesar salad. If you don't have a food processor, simply whisk the eggs in a bowl and add the oil slowly. This is a quick salad to put together when the croutons and dressing are made a day ahead.

3	egg yolks	*3*
1 Tbsp.	Dijon mustard	*15 mL*
1 1/2 cups	vegetable oil	*360 mL*
2 tsp.	white wine vinegar or lemon juice	*10 mL*
	salt and freshly ground black pepper to taste	
1 Tbsp.	Worcestershire sauce	*15 mL*
	few drops hot pepper sauce	
3	anchovy fillets	*3*
3	cloves garlic, chopped	*3*
1/4 cup	grated Parmesan cheese	*60 mL*
2 tsp.	chopped fresh basil	*10 mL*
1	romaine lettuce	*1*
6	slices bacon, cooked and crumbled	*6*
1 recipe	Garlic Herb Croutons	*1 recipe*
1/4 cup	grated Parmesan cheese	*60 mL*

Pulse the egg yolks and mustard in a food processor until blended. While the machine is running, add the oil a few drops at a time, gradually increasing to a slow steady stream. When the mixture begins to thicken, add the vinegar or lemon juice, salt and pepper. Continue blending until the sauce becomes pale. When the dressing has become very thick, add the Worcestershire sauce, hot pepper sauce, anchovy fillets, garlic, 1/4 cup (60 mL) Parmesan cheese and basil. Continue to process until all the ingredients are combined. If the dressing is too thick, add a little chicken stock or water, a spoonful at a time. Refrigerate in a glass jar until required. The dressing will keep for 3 days.

Wash and dry the lettuce. Tear the leaves into bite-size pieces and place in a serving bowl. Add the crumbled bacon, croutons and the remaining 1/4 cup (60 mL) Parmesan cheese. Toss with the dressing and serve immediately.

GARLIC HERB CROUTONS

2 Tbsp.	olive oil	30 mL
2 Tbsp.	butter	30 mL
1	clove garlic, finely chopped	1
1 Tbsp.	finely chopped Italian parsley	15 mL
1 Tbsp.	finely chopped basil	15 mL
6	slices French bread cut into 1-inch (2.5-cm) cubes	6

Preheat the oven to 350°F (175°C).

Place the oil and butter in a medium sauté pan over medium heat. When the butter has melted, stir in the garlic, parsley and basil. Add the bread cubes and stir until all the cubes are coated.

Place the bread cubes on a baking sheet and bake until golden brown, approximately 10 minutes. Turn so all the sides brown evenly and check often to prevent burning.

Let cool before using.

Classic Coleslaw

3	bacon strips, finely chopped (optional)	3
6 cups	shredded savoy cabbage	1.5 L
3	medium carrots, grated	3
2	green onions, very finely chopped	2
1/2 cup	raisins	120 mL
1 cup	Homemade Mayonnaise (page 214), or bottled salad dressing	240 mL
2 Tbsp.	cider vinegar	30 mL
	salt and freshly ground black pepper to taste	

Serves 10

This coleslaw is one of those hand-me-down recipes that turn up at family gatherings time and time again. It's very quick to make and is even better served the next day. There is a debate in our house as to whether the ingredients should be chopped by hand or shredded in a food processor. You decide.

If using bacon, sauté it until crisp in a small sauté pan over medium heat. Remove from the pan, drain on paper towel and set aside to cool.

Combine the cabbage, carrots, onion, raisins and the cooled bacon, if desired.

Whisk together the mayonnaise and cider vinegar. Season with salt and pepper. Combine the dressing and the cabbage mixture, adjusting the seasoning with salt and pepper, and mix well. Refrigerate before serving.

SALAD GREENS WITH BLUE CHEESE DRESSING

1 cup	blue cheese, crumbled	240 mL
1 cup	sour cream	240 mL
1/4 cup	mayonnaise	60 mL
1 tsp.	garlic, finely minced	5 mL
1 Tbsp.	red wine vinegar	15 mL
	salt and freshly ground black pepper to taste	
6 cups	assorted salad greens	1.5 L
1/2	red onion, finely sliced into rings	1/2
1 cup	peeled and shredded carrots	240 mL
1/2 cup	raw sunflower seeds	120 mL
1/2 cup	raisins	120 mL
3	hard-boiled eggs, chopped	3

Serves 6

This dressing is really thick, the way a classic blue cheese dressing should be. It can also be used as a dip for raw vegetables or chicken wings.

To make the dressing combine the blue cheese, sour cream, mayonnaise, garlic, vinegar, salt and pepper in a bowl and mix well. If it's too thick, thin it with a little milk. The dressing can be made up to a few days ahead and refrigerated until needed.

To make the salad, arrange the salad greens on individual serving plates. Top with a few onion slices on each plate. Place a little of the shredded carrot in the center. Sprinkle with sunflower seeds and raisins and finish with chopped egg in the center of each salad. Pass the dressing separately and allow guests to help themselves.

CLEMENTINE AND ROMAINE SALAD WITH POPPY SEED DRESSING

Serves 8

Clementines are a tiny, tangy variety of mandarin orange. You can substitute common mandarins or, for a short cut, use a tin of mandarin oranges, drained.

2/3 cup	sugar	160 mL
1/3 cup	honey	80 mL
3 Tbsp.	lemon juice	45 mL
3 Tbsp.	cider vinegar	45 mL
2 tsp.	shallot, finely chopped	10 mL
1 tsp.	dry mustard	5 mL
1/2 tsp.	salt	2.5 mL
1 cup	vegetable oil	240 mL
1 Tbsp.	poppy seeds	15 mL
2 Tbsp.	sugar	30 mL
1/2 cup	almonds, slivered	120 mL
3	clementines, peeled and sectioned	3
1	romaine lettuce	1

Combine the 2/3 cup (160 mL) sugar, honey, lemon juice, vinegar, shallot, mustard and salt in a blender. Process on high speed. While the motor is running, add the vegetable oil in a thin stream. Transfer the dressing to a bowl, and stir in the poppy seeds. Chill while preparing the rest of the ingredients.

Place the 2 Tbsp. (30 mL) sugar in a small saucepan over medium heat. Add the almonds and stir until the almonds are coated with the sugar and no sugar remains on the bottom of the pan. Set aside to cool.

Peel the clementines, removing as much of the membrane as possible. Divide them into sections. Tear the romaine leaves into bite-size pieces.

Place the romaine on individual serving plates, arrange a few clementine sections over the romaine, and sprinkle a few almonds on top. Pour a little dressing over the salad and serve immediately.

Note: For a hotter version of this dressing, add 1 tsp. (5 mL) of crushed chili peppers. The recipe makes 2 cups (480 mL) of dressing. It will keep up to 1 week in the refrigerator. Use leftover dressing as a dip for fresh fruit.

Cucumber Yogurt Salad with Maple and Dill

2	*English cucumbers, peeled and cut into 1/2-inch (1.2-cm) cubes*	2
4	*shallots, finely chopped*	4
1 Tbsp.	*chopped fresh dill*	15 mL
1/3 cup	*whipping cream*	80 mL
1/4 cup	*plain yogurt*	60 mL
1/4 cup	*apple cider vinegar*	60 mL
2 Tbsp.	*maple syrup*	30 mL
	salt and freshly ground black pepper to taste	
	sprigs of fresh dill	

Serves 4 to 6

This salad is so simple to make and completely refreshing on a hot summer's day. It pairs perfectly with grilled salmon.

Combine the cucumbers, shallots and dill in a bowl.

Whip the cream just until soft peaks form. Add the yogurt, vinegar, maple syrup, salt and pepper and mix thoroughly. Add the chopped cucumber mixture and stir well.

Cover and refrigerate. This salad is even better if left overnight. Just before serving, garnish with a few sprigs of fresh dill.

Warm Mushroom Salad with Hazelnuts

Serves 4

A warm salad is a great beginning to a winter meal and this one is particularly satisfying. You can add a few reconstituted wild mushrooms for a rich woodsy flavor.

2 Tbsp.	red wine or balsamic vinegar	30 mL
	salt and freshly ground black pepper to taste	
	pinch brown sugar	
4 Tbsp.	extra virgin olive oil	60 mL
1 Tbsp.	chopped fresh tarragon	15 mL
1	clove garlic, finely chopped	1
2 Tbsp.	extra virgin olive oil	30 mL
8 oz.	mushrooms, sliced	225 g
6 cups	lettuce greens	1.5 L
1/4 cup	hazelnuts, toasted (see page 183), skinned and chopped	60 mL

Whisk the vinegar, salt and pepper and brown sugar together until well incorporated. Mix in the 4 Tbsp. (60 mL) olive oil, tarragon and garlic.

Heat the remaining 2 Tbsp. (30 mL) olive oil in a medium sauté pan over medium heat. Add the mushrooms and sauté until wilted, approximately 5 minutes. Do not brown.

Place the lettuce in a large bowl and toss with half the vinaigrette. Arrange on serving plates.

Just before serving, remove the mushrooms from the heat. Mix in the hazelnuts and the remaining salad dressing. Spoon the mushroom mixture over the lettuce and serve immediately.

MEDITERRANEAN PASTA SALAD

2 cups	dried tricolor fusilli or penne	480 mL
1	19-oz. (540-mL) can chick peas, drained	1
10	cherry tomatoes, halved	10
1/4 cup	chopped red onion	60 mL
1	sweet green pepper, coarsely chopped	1
1	sweet red pepper, coarsely chopped	1
1 cup	English cucumber, chopped into bite-size pieces	240 mL
1/4 cup	red wine vinegar	60 mL
1 Tbsp.	lemon juice	15 mL
2	cloves garlic, finely chopped	2
2 tsp.	dried oregano	10 mL
	salt and freshly ground black pepper to taste	
2/3 cup	extra virgin olive oil	160 mL
1/2 cup	black olives	120 mL
1/2 cup	crumbled feta cheese	120 mL

Serves 6

This salad can be made a day ahead—chilling it gives the flavors time to come together. Reserve a little of the dressing when you toss the salad, as pasta tends to absorb it. Add the reserved dressing just before serving.

Bring a large pot of salted water to a boil. Add the pasta and cook until just tender (al dente). Drain, and rinse under cold running water until cooled.

In a large bowl combine the pasta, chick peas, tomatoes, red onion, green and red pepper and cucumber. Toss lightly.

In a small bowl whisk together the vinegar, lemon juice, garlic, oregano, salt and pepper. Add the olive oil and whisk until combined.

Toss the vinaigrette with the pasta and vegetables. Transfer to a serving bowl. Sprinkle the olives and feta over the top. Cover and refrigerate for at least 2 hours and up to 2 days.

SPICY ORIENTAL NOODLE SALAD

Serves 12

This spicy and light-tasting noodle salad is perfect alongside chicken brochettes or all on its own as a vegetable entrée.

1/3 cup	soy sauce	80 mL
1/4 cup	lime juice	60 mL
1 Tbsp.	minced fresh ginger	15 mL
1	clove garlic, minced	1
1/4 cup	tahini	60 mL
2 Tbsp.	honey	30 mL
1 Tbsp.	hoisin sauce	15 mL
1 tsp.	crushed red pepper flakes	5 mL
1 tsp.	sesame oil	5 mL
1	12-oz. (340-g) package oriental rice noodles	1
2 Tbsp.	vegetable oil	30 mL
1 cup	snow peas, trimmed	240 mL
2	carrots, finely julienned	2
4	green onions, sliced thinly on the diagonal	4
1	sweet red pepper, finely julienned	1
	salt and freshly ground black pepper to taste	
1/2 cup	unsalted roasted peanuts, roughly chopped	120 mL

In a medium bowl whisk together the soy sauce, lime juice, ginger, garlic, tahini, honey, hoisin sauce, red pepper flakes and sesame oil. Set aside.

Bring a large pot of salted water to a boil. Remove from the heat. Add the noodles and soak them just until they are tender but still firm to the bite, approximately 5 minutes. Drain the noodles, toss them with the vegetable oil and set them aside to cool.

Bring a small pot of salted water to a boil. Add the snow peas and cook for 1 minute. Drain and immerse in cold water until the peas are no longer warm. Drain and dry on paper towel.

Combine the noodles, snow peas, carrots, green onions and sweet red pepper in a serving bowl. Toss with the dressing and season with salt and pepper.

Just before serving, sprinkle the salad with peanuts.

BLOOD ORANGES AND FRISÉE WITH CITRUS VINAIGRETTE

1	large frisée or arugula	1
3	blood oranges	3
1/2	small red onion, thinly sliced	1/2
1/2 cup	pecans, toasted (see page 183)	120 mL
1/4 cup	white wine vinegar	60 mL
1/4 cup	frozen orange juice concentrate	60 mL
1 tsp.	grated lemon zest	5 mL
1 tsp.	grated orange zest	5 mL
	salt and freshly ground black pepper to taste	
	pinch ground ginger	
3/4 cup	vegetable oil	180 mL

Serves 6

Blood oranges with their ruby-red color make a nice change. If blood oranges are not available substitute navel oranges.

Wash the frisée well and tear it into bite-size pieces. Arrange on individual serving plates.

With a sharp knife, remove the skin and pith of the oranges. Section the oranges, and remove the tough membranes. Arrange the orange sections and onion slices over the frisée pieces. Sprinkle with toasted pecans.

In a bowl whisk together the vinegar, orange juice concentrate, lemon and orange zest, salt, pepper and ginger. Add the vegetable oil and whisk again.

Just before serving, pour a little vinaigrette over each salad, or pass it around and let guests help themselves.

PECANS WITH BRIE OVER BABY GREENS AND TROPICAL MANGO VINAIGRETTE

Serves 6 to 8

This is a lively salad to make on a warm summer's night. The sweetness of mango is a perfect foil for the tart vinegar. If you don't have maple syrup, substitute brown sugar.

1/2 cup	mango, peeled and cubed	120 mL
1	small shallot, peeled	1
1/4 cup	white wine vinegar	60 mL
1/2 cup	vegetable oil	120 mL
2 Tbsp.	maple syrup	30 mL
1/4 tsp.	salt	1.2 mL
1/4 tsp.	freshly ground black pepper	1.2 mL
1/8 tsp.	curry powder	.6 mL
1	large red leaf lettuce	1
1	endive	1
8 cups	mesclun mix (mixed baby greens)	2 L
1	large mango, peeled and sliced	1
6 oz.	Brie cheese, cut into thin wedges	170 g
1/2 cup	whole pecans, lightly toasted (see page 183)	120 mL

Make the vinaigrette before the salad to give the flavors time to come together before serving. Place the mango, shallot, vinegar, oil, maple syrup, salt, pepper and curry powder in a blender or food processor. Blend until the mango and shallot are puréed. The dressing can be made a day ahead and stored in the refrigerator.

Arrange the salad on individual serving plates for a really special presentation. For a buffet, arrange it on a large serving platter.

Wash all the vegetables thoroughly and spin or pat them dry. Just before serving, line up the salad plates and place one or two large leaves of lettuce on each plate. Place two leaves of Belgian endive on top of the lettuce. Mound a large cupful of baby greens on top. Place a few slices of mango on top and then wedges of Brie. Sprinkle with a few pecans. Drizzle with a few spoonfuls of salad dressing and serve immediately.

Antipasto (page 12)

Red Pepper and Chutney Phyllo Triangles
(page 26)

Savory Pepper and Mushroom Profiteroles
(page 28)

Potato, Cheddar and Beer Soup
(page 62)

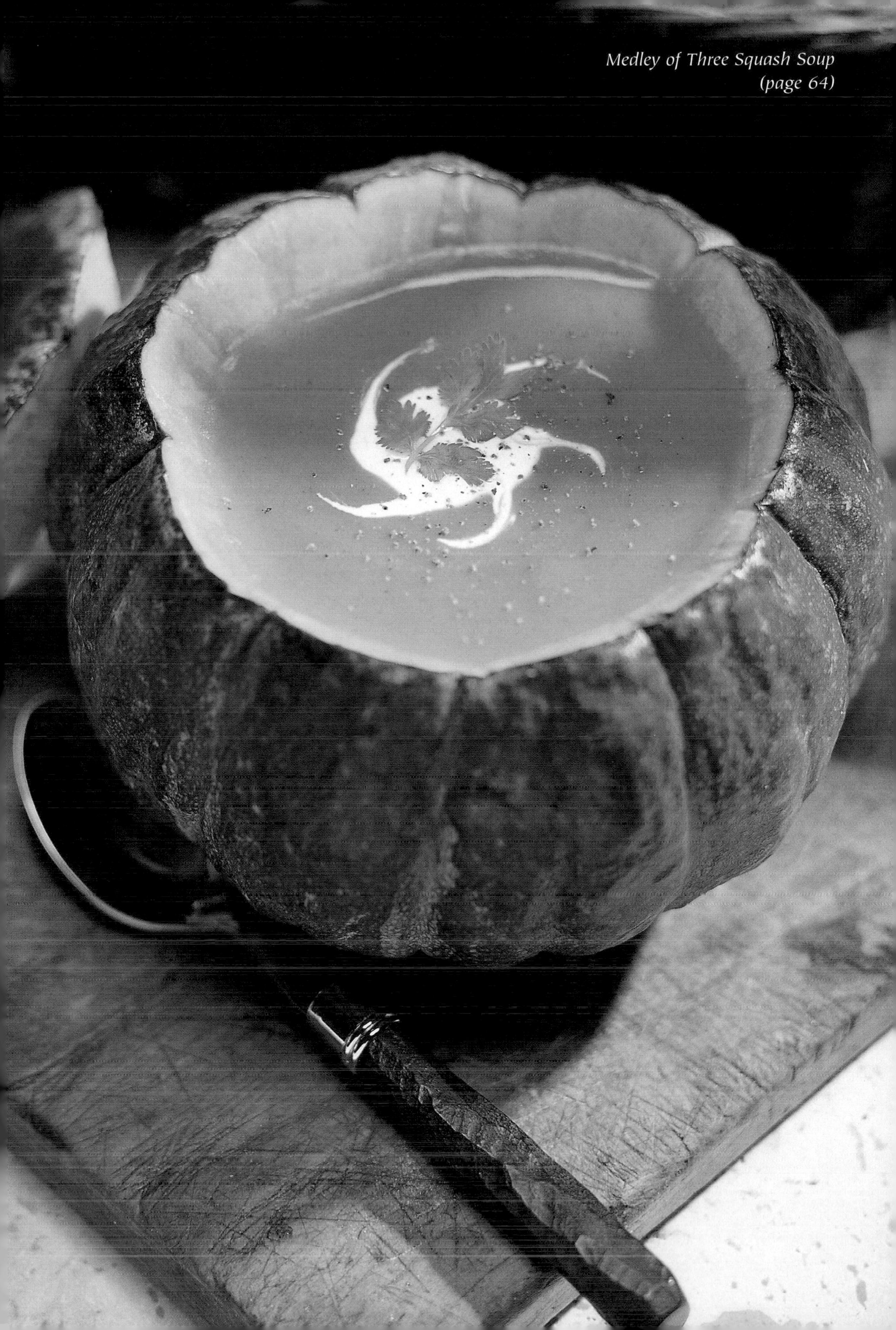

Medley of Three Squash Soup
(page 64)

Roasted Beet Salad with Fresh Pears and Walnuts over Baby Greens (page 74)

Broccoli Salad with Sunflower Seeds
(page 77)

Cauliflower and Potato Curry
(page 107)

Spinach Salad with Sun-Dried Cranberries and Cranberry Vinaigrette

4 cups	fresh spinach leaves, loosely packed	1 L
1/2 cup	slivered almonds, toasted (see page 183)	120 mL
1/3 cup	sun-dried cranberries	80 mL
1/2	small red onion, very thinly sliced	1/2
2 Tbsp.	cider vinegar	30 mL
2 Tbsp.	cranberry juice	30 mL
1 Tbsp.	maple syrup	30 mL
1/2 tsp.	Dijon mustard	2.5 mL
	salt and freshly ground black pepper to taste	
1/2 cup	vegetable oil	120 mL

Serves 4

This festive-looking salad makes a nice opener for Christmas dinner.

Wash and dry the spinach leaves. Place in a large serving bowl. Place the almonds, cranberries and onion rings on top of the spinach.

Place the vinegar, cranberry juice, maple syrup, Dijon mustard, salt and pepper in a jar with a tight-fitting lid. Shake or whisk until well combined. Add the vegetable oil and shake again. Just before serving, shake the vinaigrette, pour it over the salad and serve.

OLD-FASHIONED POTATO SALAD

Serves 6 to 8

Just like mom used to make. Garnish the top with a few slices of hard-boiled egg and a sprinkling of paprika for a traditional look. The dressing is an old-fashioned boiled dressing that can be prepared ahead and stored in a glass jar in the refrigerator for up to 1 week.

For the dressing:

1 tsp.	salt	5 mL
2 tsp.	dry mustard	10 mL
2 Tbsp.	brown sugar	30 mL
2 Tbsp.	all purpose flour	30 mL
	pinch cayenne pepper	
2 cups	milk	480 mL
3	egg yolks	3
3 Tbsp.	butter	45 mL
1/3 cup	cider vinegar	80 mL
	freshly ground black pepper to taste	
1 Tbsp.	Dijon mustard (optional)	15 mL

To make the dressing, combine the 1 tsp. (5 mL) salt, dry mustard, sugar, flour and cayenne in the top of a double boiler. Combine the milk and egg yolks. Gradually add the milk mixture to the dry mixture, stirring constantly over medium heat until the mixture is thick. Stir in the butter and vinegar. Remove from the heat. Season with pepper. If you like a mustard style, add the Dijon mustard. Chill the dressing.

To make the salad:

3 lbs.	new potatoes, peeled and roughly chopped	1.4 kg
6	hard-boiled eggs, chopped	6
6	green onions, chopped	6
1	sweet red pepper, finely chopped	1
4	radishes, finely chopped	4
1/2 cup	finely chopped celery	120 mL
1/2	sweet green pepper, finely chopped	1/2
3/4 tsp.	sweet paprika	4 mL
	salt and black pepper to taste	

Place the potatoes in a large pot of cold salted water over high heat. Bring to a boil and cook until tender. Drain and set aside to cool.

Chop the potatoes into small pieces. Place them in a large bowl with the chopped eggs, green onions, red pepper, radishes, celery and green pepper. Sprinkle with the paprika, salt and pepper and toss lightly. Pour the dressing over the vegetables and mix well. Chill before serving.

HOT RADICCHIO AND GORGONZOLA SALAD

2 Tbsp.	butter	30 mL
2	heads radicchio, cut in half	2
6 oz.	Gorgonzola cheese	170 g
2 Tbsp.	balsamic vinegar	30 mL
1 Tbsp.	butter	15 mL
1 Tbsp.	chopped fresh basil	15 mL
	salt and freshly ground black pepper to taste	

Serves 4

Melt the 2 Tbsp. (30 mL) of butter in a medium sauté pan over medium heat. Add the radicchio cut side down and sauté for 4 to 5 minutes. Turn it and cook the other side until the leaves have wilted. Transfer the radicchio to a baking sheet, cut side up. Distribute the Gorgonzola over the radicchio and place it under the broiler until the cheese has melted.

Meanwhile, return the sauté pan to the heat. Add the vinegar, the 1 Tbsp. (15 mL) of butter and the basil. Gently shake the pan over medium heat until the butter has melted, about 1 minute.

Arrange the radicchio on a serving plate. Spoon the vinegar mixture over the radicchio, season with salt and pepper and serve immediately.

A beautiful Italian salad featuring all the best from Italy: Gorgonzola cheese, balsamic vinegar and radicchio. If Gorgonzola is not available, try another type of blue cheese—not as authentic but equally as good. This salad also makes a nice side dish.

CURRIED RICE SALAD

Serves 6

This has long been a favorite. Add poached and diced chicken breast to the salad and make it a meal. This salad can be prepared a day ahead.

2 Tbsp.	vegetable oil	30 mL
1	medium onion, finely chopped	1
1 Tbsp.	curry powder	15 mL
1/2 tsp.	finely chopped fresh ginger	2.5 mL
1 cup	rice	240 mL
1 tsp.	salt	5 mL
2 cups	water	480 mL
1 cup	unsweetened coconut	240 mL
2	stalks celery, finely chopped	2
2	medium sweet red peppers, cut into 1/4-inch (.6-cm) dice	2
3/4 cup	raisins	180 mL
1	Granny Smith apple, cored and cut into 1/4-inch (.6-cm) dice	1
1/2 cup	red wine vinegar	120 mL
1 Tbsp.	lemon juice	15 mL
1 Tbsp.	honey	15 mL
1 Tbsp.	Dijon mustard	15 mL
	pinch salt	
2 tsp.	freshly ground black pepper	10 mL
1 Tbsp.	garlic, minced	15 mL
6 Tbsp.	vegetable oil	90 mL

Heat the 2 Tbsp. (30 mL) of oil in a large saucepan over medium-high heat. Add the onion and sauté for 2 minutes. Add the curry powder and ginger and sauté for 30 seconds. Add the rice and sauté for 1 minute, stirring well. Add the salt and water, cover, and bring to a boil. Reduce the heat to low and simmer for 15 minutes. Be careful the rice does not boil dry. Let sit 5 minutes, remove the lid and fluff with a fork. Set aside to cool.

Toast the coconut in the oven or in a skillet over medium heat until it's lightly browned.

In a large bowl combine the toasted coconut, cooled rice, celery, red pepper, raisins and apple.

Combine the vinegar, lemon juice, honey, mustard, salt, pepper and garlic in a bowl. Whisk in the 6 Tbsp. (90 mL) of vegetable oil until well combined. Pour the dressing over the rice mixture. Cover and chill well before serving.

RASPBERRY AND TARRAGON SALAD WITH RASPBERRY VINAIGRETTE

2	*heads Boston lettuce*	2
2 Tbsp.	*fresh tarragon*	30 mL
1/4 cup	*slivered almonds*	60 mL
1/2 cup	*fresh or frozen raspberries*	120 mL
3 Tbsp.	*honey*	45 mL
1/4 cup	*white wine or champagne vinegar*	60 mL
1/2 cup	*vegetable oil*	120 mL
2 tsp.	*fresh tarragon*	10 mL
	salt and freshly ground black pepper to taste	
1 cup	*fresh raspberries*	240 mL

Serves 6

This salad is completely irresistible when summer's berries are at their peak.

Remove and discard the outer layers of the lettuce. Tear the tender inner leaves into bite-size pieces. Toss with the tarragon and almonds.

To make the vinaigrette, place the 1/2 cup (120 mL) raspberries, honey, vinegar, oil and 2 tsp. (10 mL) tarragon in a food processor. Pulse on low speed just until combined. Season with salt and pepper.

Just before serving, toss the salad with as much of the vinaigrette as desired and scatter the fresh raspberries on top.

Spinach, Tomato and Asparagus Salad with Sun-Dried Tomato Vinaigrette

Serves 6

A trip to Tuscany could have inspired this flavorful salad, which features a number of ingredients synonymous with northern Italy. A great opener for your next pasta night.

2	sun-dried tomato halves	2
3 Tbsp.	balsamic vinegar	45 mL
1	small clove garlic, minced	1
	salt and freshly ground black pepper to taste	
1/3 cup	extra virgin olive oil	80 mL
1 Tbsp.	finely chopped fresh basil	15 mL
12	slices prosciutto	12
6	plum tomatoes, halved	6
1	clove garlic, halved	1
2 Tbsp.	olive oil	30 mL
1/2 lb.	asparagus	225 g
1/2 lb.	baby spinach, washed and dried	225 g
1/2 cup	shaved Parmesan cheese	120 mL

Bring a small saucepan of water to a boil. Add the sun-dried tomatoes and simmer, just covered with water, for 3 minutes, or until tender. Drain and chop them very fine.

Combine the tomatoes, vinegar, garlic, salt and pepper in a bowl. Whisk in the oil in a slow steady stream. Stir in the basil. Set the dressing aside.

Preheat the oven to 350°F (175°C).

Place the prosciutto slices and the halved fresh tomatoes, cut side up, on a baking sheet. Rub the cut side of the tomatoes with the garlic clove, drizzle with olive oil and sprinkle with salt and pepper. Bake for 20 minutes, or until the prosciutto is crisp and the tomatoes are soft.

Bring a large saucepan of lightly salted water to a boil. Add the asparagus and cook until just tender, 3 to 4 minutes. Drain and plunge the asparagus immediately into cold water to halt the cooking process. Drain and pat dry on paper towel. Cut each spear in half lengthwise and set aside.

Arrange the spinach and asparagus on individual serving plates. Top with the tomatoes, prosciutto and shaved Parmesan. Drizzle the dressing over top. Serve the salad while the tomatoes and prosciutto are still warm.

TABOULEH

1 cup	bulgur (or cracked wheat)	240 mL
1 cup	finely chopped fresh parsley	240 mL
1 cup	finely chopped green onion	240 mL
3/4 cup	finely chopped fresh mint	180 mL
1	sweet red pepper, finely chopped	1
3	tomatoes, seeded and finely chopped	3
1/3 cup	lemon juice	80 mL
	salt and freshly ground black pepper to taste	
1/3 cup	extra virgin olive oil	80 mL
	sprigs of fresh mint	

Serves 6

A flavorful salad from the Middle East. Serve Tabouleh alongside lamb or beef kebabs.

Bring a small pot of lightly salted water to a boil. Add the bulgur and let sit for 10 minutes, or until it begins to plump up. Drain well.

In a medium bowl combine the bulgur, parsley, onion, mint, red pepper and tomatoes.

Combine the lemon juice, salt and pepper, stirring until the salt is dissolved. Add the olive oil. Stir well and pour over the bulgur mixture. Mix well. Refrigerate until ready to serve.

Garnish with fresh mint sprigs.

Spinach Salad with Warm Bacon Dressing

6 cups	fresh spinach leaves, stems removed	1.5 L
8	slices bacon, chopped	8
1 Tbsp.	cider vinegar	15 mL
2 tsp.	brown sugar	10 mL
	salt and freshly ground black pepper to taste	
1/3 cup	sour cream	80 mL
2	hard-boiled eggs, chopped	2
1/4 cup	raisins	60 mL
1/4 cup	walnut pieces, toasted (see page 183)	60 mL

Serves 6

This is a time-honored favorite, brought out of the cupboard and dusted off for old friends. A warm dressing on a salad is unusual and a welcome change.

Wash the spinach well. Make sure no sand remains on the leaves. Spin or pat dry.

Fry the bacon in a medium sauté pan over medium heat until it's just beginning to crisp. Drain on paper towel and set aside. Reserve 4 Tbsp. (60 mL) of the bacon drippings.

Return the pan and the reserved bacon drippings to the heat. In a small bowl whisk together the cider vinegar, brown sugar, salt and pepper. Add the mixture slowly to the warm bacon drippings. Stir well. Cool slightly before adding the sour cream.

Place the spinach, bacon pieces, eggs, raisins and walnut pieces in a large bowl. Toss all the ingredients with the dressing and serve immediately.

SIDE DISHES

Deciding what vegetable to serve and how to make it a memorable part of the meal is a question that plagues all cooks planning a menu. Vegetable side dishes can become a lot more tempting when just a little forethought goes into the dish.

Some of the following dishes take a little more preparation time, but are well worth the effort. Others are fast and simple and use ingredients found in everyone's pantry.

Some call for vegetables that are relatively new to our grocery stores, such as Jerusalem artichokes and salsify, but both have been around for a very long time. If wondering how to prepare them has prevented you from bringing them home, now is the time to let these recipes introduce you to these interesting and delicious vegetables.

GRILLING VEGETABLES

Grilling is quickly becoming one of the most popular—and certainly tastiest—methods of preparing vegetables. When grilling, always cook over medium heat with the lid down, turning the vegetables often. Spiced grilling oil for brushing on the vegetables can be found on page 215. Here are some cooking times and suggestions for bastes to use when grilling vegetables.

Vegetable	Preparation	Baste	Grill Time (minutes)
carrots	cut carrots 1/4 inch (.6 cm) thick, on the diagonal or lengthwise	spiced oil, or butter with cinnamon or curry powder	15–20
corn on the cob	blanch in boiling water 5 minutes, drain, and transfer to grill	spiced oil or thinned pesto	5–10
eggplant	slice 1/4 inch (.6 cm) thick	spiced oil	10–12
mushrooms	thread on skewers	spiced oil	5–7
portobello mushrooms	slice or leave whole	spiced oil or thinned pesto	10–15
onions	slice 1/4 inch (.6 cm) thick into wedges	spiced oil	15–18
peppers	slice into 6 or 8 wedges, discard seeds	spiced oil or thinned pesto	10–12
potatoes	slice 1/4 inch (.6 cm) thick	spiced oil	10–15
squash (winter)	slice 1/2 inch (1.2 cm) thick at the widest part	butter with curry powder	10–15
sweet potatoes	slice 1/4 inch (.6 cm) thick	spiced oil, or butter with cinnamon	10–15
tomatoes	cut in half	spiced oil or thinned pesto	5–6
cherry tomatoes	thread on skewers	spiced oil or thinned pesto	5–6
zucchini	slice 1/4 inch (.6 cm) thick on the diagonal	spiced oil or thinned pesto	6–8

Asparagus Risotto

2 Tbsp.	oil	30 mL
2 Tbsp.	butter	30 mL
1	small onion, finely chopped	1
1 cup	Arborio rice	240 mL
1 cup	white wine	240 mL
4 cups	chicken stock	1 L
1 lb.	asparagus spears	455 g
2 Tbsp.	butter	30 mL
1/2 cup	Parmesan cheese	120 mL

Heat the oil and melt 2 Tbsp. (30 mL) of butter in a medium sauté pan over medium heat. Add the onion and sauté for 1 or 2 minutes, until it begins to soften. Add the rice and sauté over low heat for 1 to 2 minutes. Add the wine and stir until it has evaporated. Add the stock slowly, about 1/3 cup (80 mL) at a time, and continue to cook, stirring constantly, until all the stock is absorbed and the rice is tender, approximately 20 minutes. If it gets a little dry before the rice is cooked, add more stock.

Chop the asparagus spears into 2- to 3-inch (5- to 7.5-cm) lengths. Reserve the tips. Wrap the remainder of the spears in plastic wrap and freeze for making soup later. Just before the risotto is ready, bring a pot of lightly salted water to a boil and add the asparagus tips. Cook for 3 to 4 minutes, until tender. Drain. Set aside.

Add the remaining 2 Tbsp. (30 mL) of butter to the risotto. Stir in the Parmesan cheese and asparagus tips. Serve hot. Pass around extra grated Parmesan cheese for guests to help themselves.

Serves 2

This is my favorite way to serve asparagus. Sometimes we have it as a main dish, sometimes as a side dish. The creamy texture of risotto is achieved by stirring constantly while it's cooking.

ASPARAGUS FLANS

2 lbs.	*fresh asparagus*	900 g
2 Tbsp.	*whipping cream*	30 mL
1/2 tsp.	*dried tarragon*	1.2 mL
3 Tbsp.	*butter, softened*	45 mL
1/4 cup	*grated Parmesan cheese*	60 mL
1/2 tsp.	*salt*	2.5 mL
3	*large eggs*	3
1 Tbsp.	*butter*	15 mL

Serves 6

If you like, make these little flans from just the stalks of the asparagus, saving the tender tips for steaming on their own.

Preheat the oven to 350°F (175°C).

Butter six 3/4-cup (180-mL) ramekins and line the bottoms with parchment paper. Place a baking rack in a baking pan large enough to hold the ramekins.

Trim the asparagus tips to 2-inch (5-cm) lengths and set aside.

Cut the stalks into 1-inch (2.5-cm) pieces. Bring a pot of lightly salted water to a boil and add the asparagus pieces. Cook until just tender, about 3 to 4 minutes. Transfer the asparagus to a cold water bath to stop the cooking process. Drain and pat dry with paper towel.

Place the asparagus stalks, cream, tarragon, 3 Tbsp. (45 mL) butter, Parmesan cheese and salt in a blender or food processor and purée until smooth.

Whisk the eggs in a bowl until well combined and add the asparagus purée in a stream, whisking until smooth.

Divide the mixture among the ramekins and place on the rack inside the baking pan. Add very hot water until it comes halfway up the sides of the ramekins. Bake in the lower third of the oven for 35 to 40 minutes, or until a knife inserted in the center comes out clean. Remove from the oven and cool for 5 minutes.

Bring a small saucepan of salted water to a boil. Add 6 of the reserved asparagus tips and cook until just tender. Plunge into cold water. Drain and pat dry.

Just before serving, slice each asparagus tip in half lengthwise. Melt the remaining 1 Tbsp. (15 mL) of butter, add the asparagus tips and cook until they are just heated through. Remove the flans from the ramekins. Place 2 halves of asparagus tip on each flan.

STEAMED BROCCOLI WITH MUSTARD SAUCE

3 lbs.	broccoli, florets only	1.4 kg
2 Tbsp.	sugar	30 mL
1/4 cup	white wine vinegar	60 mL
1/2 cup	Dijon mustard	120 mL
2/3 cup	vegetable oil	160 mL

Serves 6

Bring a few inches of water to a boil in a vegetable steamer. Add the broccoli and cook until just barely tender, approximately 10 minutes. Transfer to a warmed serving dish.

While the broccoli is cooking, whisk the sugar and vinegar until the sugar has dissolved. Whisk in the mustard. Add the oil slowly and continue to whisk until well blended. Thin with a little water if the sauce is too thick. Pour the sauce over the hot vegetables just before serving.

The sauce can be made a day ahead and stored in the refrigerator; bring it to room temperature before serving.

Steamed broccoli retains its bright color and steaming gives you better control of the cooking process, eliminating that mushy broccoli problem. This mustard sauce is also delicious with roasted beets or steamed green beans.

Broccoli and Almond Gratin

1 lb.	fresh broccoli florets, cut into large pieces, stems cut into bite-size pieces	455 g
1/4 cup	butter	60 mL
2	medium onions, thinly sliced	2
1/2 cup	slivered almonds, lightly toasted (see page 183)	120 mL
2 Tbsp.	butter	30 mL
2 Tbsp.	flour	30 mL
1 cup	milk	240 mL
1/4 tsp.	freshly ground black pepper	1.2 mL
1/4 tsp.	dry mustard	1.2 mL
4 oz.	cream cheese, cubed	113 g
1/2 cup	grated Cheddar cheese	120 mL
1/2 cup	fresh bread crumbs	120 mL
1/2 cup	almonds, finely chopped	120 mL
2 Tbsp.	butter, melted	30 mL

Serves 6

This hearty casserole is so rich it could almost be the main course. Serve it with roasted meat or poultry.

Bring a pot of lightly salted water to a boil. Add the broccoli and cook until barely tender. Remove and plunge into cold water to halt the cooking process. Drain and place in a bowl.

Melt the 1/4 cup (60 mL) butter in a medium sauté pan over medium-high heat. Sauté the onion until very soft, 5 to 8 minutes. Add the onion to the broccoli along with the toasted slivered almonds.

Preheat the oven to 350°F (175°C).

Melt the 2 Tbsp. (30 mL) of butter in a small saucepan over medium heat. When it's bubbly, stir in the flour until smooth. Slowly add the milk, continuing to cook until smooth and thickened. Add the black pepper, dry mustard and cheeses. Continue to stir until the cheese has melted. Remove from the heat.

Butter an 8- x 8-inch (20- x 20-cm) baking dish. Place the broccoli mixture in the baking dish. Pour the cheese mixture over top.

Cover the casserole with aluminum foil. Line a baking sheet with aluminum foil in case the casserole boils over. Place the casserole on the baking sheet and bake for 20 to 30 minutes or until heated through.

Combine the bread crumbs, chopped almonds and 2 Tbsp. (30 mL) of melted butter in a bowl. Toss until well combined. Remove the casserole from the oven and top with the bread crumb mixture. Continue baking until the crumbs are golden brown, 5 to 10 minutes. Let the casserole rest 10 minutes before serving.

Note: The casserole can be assembled, refrigerated overnight and baked the next day. Bake an extra 10 minutes before adding the bread crumbs.

BRUSSELS SPROUTS IN BALSAMIC VINEGAR

1 lb.	Brussels sprouts	455 g
1 Tbsp.	butter	15 mL
2 Tbsp.	balsamic vinegar	30 mL
1/2 tsp.	sugar	2.5 mL

Serves 4

This recipe for Brussels sprouts brings out their best flavor.

Remove the tough stems and any wilted outer leaves from the Brussels sprouts. Cut an X in the bottom of each sprout. This ensures even cooking. Steam the Brussels sprouts until they are just tender.

Melt the butter in a pot over low heat. Add the balsamic vinegar and sugar. Stir until the sugar is dissolved. Add the sprouts and stir until they are well coated with the vinegar mixture. Serve immediately.

Brussels Sprouts with Bacon

Serves 4 to 5

The smoky taste of bacon is the basis for this delicious sauce.

1	egg, lightly beaten	1
2 Tbsp.	sugar	30 mL
1/3 cup	white vinegar	80 mL
3 Tbsp.	water	45 mL
1/4 tsp.	dry mustard	1.2 mL
	salt and freshly ground black pepper to taste	
6	slices bacon, finely chopped	6
1 lb.	Brussels sprouts	455 g

Whisk together the egg, sugar, vinegar, water, mustard, salt and pepper. Cover and refrigerate.

Sauté the bacon until crisp in a skillet over medium-high heat. Drain the bacon on paper towel. Reserve 2 Tbsp. (30 mL) of the bacon drippings and discard the rest. Reduce the heat to low. Stirring constantly, add the egg mixture to the pan. Continue stirring until the sauce has thickened. Set aside.

Bring a saucepan of well-salted water to a boil. Cut an X in the bottom of each Brussels sprout to ensure even cooking. Cook for approximately 8 minutes, or until the sprouts are just tender. Drain the sprouts. Toss them with the sauce and transfer to a serving dish. Garnish with chopped bacon and serve immediately.

Red Cabbage with Juniper

2 Tbsp.	butter	30 mL
1	large onion, finely chopped	1
6	juniper berries, lightly crushed	6
1/2	head red cabbage, cored and shredded	1/2
1/2 cup	vegetable stock	120 mL
1 Tbsp.	red currant jelly	15 mL
1 Tbsp.	red wine vinegar	15 mL
	salt and freshly ground black pepper to taste	
1	apple, peeled, cored and shredded	1

Serves 4 to 6

The sweet-sour burst of flavor in this red cabbage dish pairs particularly well with pork or game.

Heat the butter in a medium sauté pan over medium-high heat. Add the onion and sauté until translucent. Add the juniper berries and sauté until they release their aroma.

Add the cabbage, stock, red currant jelly and vinegar. Bring to a boil. Season with salt and pepper. Cover and simmer for 15 to 20 minutes, or until the cabbage is tender.

Add the apple, cover, and cook a further 10 minutes. Adjust the seasoning. Serve hot.

CARROTS VICHY

2 cups	carrots, peeled and thinly sliced on the diagonal	480 mL
1/2 cup	Vichy water, boiling	120 mL
2 Tbsp.	butter	30 mL
1 Tbsp.	sugar	15 mL
1/4 tsp.	salt	1.2 mL
1 tsp.	lemon juice	5 mL
1/2 tsp.	grated lemon zest	2.5 mL
	fresh parsley sprigs (optional)	

Serves 4

To make authentic Carrots Vichy you must use Vichy water from France. But tap water works just as well.

Place the carrots, Vichy water, butter, sugar, salt, lemon juice and lemon zest in a saucepan with a tight-fitting lid. Cover and simmer until the carrots are tender and well glazed, approximately 10 minutes. Check to be sure they do not dry out.

Transfer to a heated serving dish and garnish with fresh parsley.

CAULIFLOWER AND POTATO CURRY

4 Tbsp.	vegetable oil	60 mL
2	large onions, finely chopped	2
4	cloves garlic, finely chopped	4
1/2 tsp.	finely chopped fresh ginger	2.5 mL
1 1/2 Tbsp.	curry powder or paste	22.5 mL
1	cinnamon stick	1
	pinch ground cloves	
1/4 tsp.	freshly ground black pepper	1.2 mL
1/2 tsp.	sweet paprika	2.5 mL
1/2 cup	water	120 mL
2	large potatoes, peeled and cut into 2-inch (5-cm) cubes	2
1 tsp.	salt	5 mL
1/2	large head cauliflower, cut into florets	1/2
1	14-oz. (398-mL) can coconut milk	1
1/3 cup	shredded unsweetened coconut	80 mL

Serves 4

Serve this lively Indian curry with rice as a main course or as a spicy complement to roast lamb.

Heat 2 Tbsp. (30 mL) of the oil in a non-stick frying pan over medium-high heat. Add the onions and sauté them until translucent. Set the onions aside.

Return the pan to the heat and add the remaining oil. Sauté the garlic and ginger for 30 seconds, then add the curry, cinnamon stick, cloves and black pepper. Sauté for 1 minute, stirring constantly.

Return the onions to the pan. Add the paprika and a little of the water to loosen the spices from the bottom of the pan. Add the potatoes and sauté for 5 minutes. Add the remainder of the water and sprinkle with the salt. Cover and cook for 6 to 7 minutes. Add the cauliflower and coconut milk and simmer until tender. Remove and discard the cinnamon stick.

Just before serving, sprinkle the shredded coconut over the top.

CELERY ROOT AND POTATOES ANNA

3/4 cup	unsalted butter, melted	180 mL
6	new potatoes, peeled and thinly sliced (about 1 1/2 lb./680 g)	6
	salt and freshly ground black pepper to taste	

Serves 6 | 2 | celery roots, peeled and thinly sliced (about 1 lb./455 g) | 2 |

Traditionally Potatoes Anna are made with potatoes only but we like the added flavor that celery root gives this dish.

Preheat the oven to 400°F (200°C).

Heat a 10-inch (25-cm) cast-iron frying pan over medium heat. Add a few spoonfuls of the butter. Working quickly, line the pan with a layer of potatoes, starting from the center. Sprinkle with salt and pepper, and drizzle with butter. Add a layer of celery root, season with salt and pepper, and drizzle with a little more butter. Shake the pan to be sure the bottom layer is not sticking. Continue alternating layers of potato and celery root, sprinkling each with salt and pepper and drizzling with butter, until all the slices are arranged.

Press the potatoes into the pan with a spatula. Cover with a tight-fitting lid or aluminum foil and bake for 20 minutes. Remove the lid, press the mixture down again and continue to bake without a lid until the vegetables are tender, approximately 20 to 30 minutes longer.

Remove the pan from the oven and let it sit a few minutes. Slide a long thin spatula around the sides of the pan and underneath the potatoes. Place a large plate over the top of the pan and quickly invert the pan onto the plate. Be careful, as hot butter may drip from the pan. Serve immediately.

CREOLE CORN SAUTÉ

2 Tbsp.	butter	30 mL
1	small onion, finely chopped	1
1/2 cup	finely chopped celery	120 mL
1	small sweet green pepper, finely diced	1
1	small sweet red pepper, finely diced	1
1	clove garlic, finely chopped	1
1	large tomato, seeded and chopped	1
1 cup	okra, trimmed into 1-inch (2.5-cm) lengths	240 mL
1 tsp.	Cajun seasoning	5 mL
4	ears corn, kernels removed, or 2 1/2 cups (600 mL) frozen	4
1/2 cup	chicken stock	120 mL
1	bay leaf	1
	salt and freshly ground black pepper to taste	

Serves 6 to 8

This piquant side dish of corn goes with just about everything. It can be prepared a day ahead and reheated just before serving.

Melt the butter in a large sauté pan over medium-high heat. Add the onion and celery and sauté for 3 minutes. Add the green and red pepper and continue to sauté until the onion and celery begin to soften, approximately 5 minutes.

Reduce the heat, add the garlic and sauté for 30 seconds. Add the tomato, okra and Cajun seasoning and continue to sauté for 10 minutes, stirring frequently.

Add the corn, chicken stock and bay leaf; continue to cook until the corn is tender, about 7 minutes. Season with salt and pepper. Discard the bay leaf and serve immediately.

SOUTHERN CORN CAKES

1 cup	cornmeal	240 mL
1/2 cup	all purpose flour	120 mL
1 tsp.	salt	5 mL
1 tsp.	baking soda	5 mL
	freshly ground black pepper to taste	
	pinch cayenne pepper	
1 tsp.	sugar	5 mL
2 Tbsp.	butter, melted	30 mL
1	large egg	1
1 cup	buttermilk	240 mL
1	ear corn, kernels removed, or 1 cup (240 mL) frozen	1
1	small onion, finely chopped	1
1/2	sweet red pepper, finely chopped	1/2
1 tsp.	crushed red chili peppers	5 mL
1 cup	grated Monterey Jack or mild Cheddar cheese	240 mL

Makes approximately 12 cakes

These corn cakes are a great alternative to potatoes alongside barbecued ribs or steaks.

In a medium bowl stir together the cornmeal, flour, salt, baking soda, black pepper, cayenne pepper and sugar.

In a separate bowl whisk together the butter, egg and buttermilk. Add the corn, onion, red pepper, chili pepper and cheese. Add the cornmeal mixture, stirring just until it's combined.

Heat a non-stick griddle over medium heat. Brush it with a little vegetable oil. Using a 1/4-cup (60-mL) measure, pour the batter onto the griddle, spreading it to make a 4- to 5-inch (10- to 12.5-cm) cake. Cook for 2 to 3 minutes, until the cake is lightly browned, then turn it and brown the other side. Keep the cakes warm in the oven until serving time.

GRATIN OF FENNEL AND PARMESAN

2	medium bulbs fennel, washed and trimmed	2
1/2 cup	chicken stock	120 mL
2 Tbsp.	olive oil	30 mL
	salt and freshly ground black pepper to taste	
1/2 cup	grated Parmesan cheese	120 mL

Preheat the oven to 400°F (200°C).

Cut each fennel bulb in half, and then cut each half into 3 wedges, leaving some core attached to each wedge. Place the wedges in a 9- x 9-inch (23- x 23-cm) baking dish. Pour the stock over the wedges, then the olive oil. Sprinkle with salt and pepper. Cover the pan with aluminum foil and bake for 20 minutes.

Remove the foil, sprinkle with the Parmesan cheese and continue baking until the fennel is tender and the cheese browned, approximately 30 minutes. Serve immediately.

Serves 4 to 6

The subtle licorice flavor of fennel has long been an Italian favorite. Serve this with roasted meats or chicken.

GREEN AND YELLOW BEANS IN BALSAMIC VINEGAR

3/4 lb.	green beans, trimmed	340 g
3/4 lb.	yellow beans, trimmed	340 g
2 Tbsp.	olive oil	30 mL
1	small onion, finely chopped	1
1	clove garlic, finely chopped	1
1/4 cup	balsamic vinegar	60 mL
1/3 cup	olive oil	80 mL
1 tsp.	Dijon mustard, with seeds if possible	10 mL
	salt and freshly ground black pepper to taste	
1/2 cup	grated Parmesan cheese	120 mL

Serves 6

These tangy beans are sensational with beef or lamb dishes.

Bring a large pot of salted water to a boil. Add the beans and cook until barely tender, about 5 to 6 minutes. Drain, rinse under cold water and set aside.

Heat the 2 Tbsp. (30 mL) olive oil in a sauté pan over medium heat. Add the onion and sauté until translucent. Add the garlic and sauté for 30 seconds. Add the vinegar and cook until the liquid is reduced by half.

Remove the pan from the heat and add the remaining olive oil. Stir in the Dijon mustard, salt and pepper. The dressing should be warm but not hot. Add the beans and Parmesan cheese and mix well.

STEWED LIMA BEANS IN GARLIC AND BROTH

2 Tbsp.	butter	30 mL
2	cloves garlic, finely chopped	2
1 lb.	fresh or frozen lima beans	455 g
1 cup	chicken stock	240 mL
	salt and freshly ground black pepper to taste	

Melt the butter in a medium saucepan over medium heat. When it bubbles, add the garlic and sauté for 30 seconds. Add the lima beans and stock, bring to a boil, reduce the heat and simmer covered for 5 minutes. Season with salt and pepper.

Serves 4 to 6

If you've been thinking lima beans are bland, try them prepared this way. Garlic gives them a real flavor boost and the broth becomes a sauce for the beans.

ROASTED JERUSALEM ARTICHOKES

2 lbs.	Jerusalem artichokes	900 g
3 Tbsp.	vegetable oil	45 mL
1 tsp.	finely chopped fresh rosemary	5 mL
	salt and freshly ground black pepper to taste	

Preheat the oven to 375°F (190°C).

Peel the artichokes and cut them in half. Toss immediately with the oil to prevent browning. Sprinkle with rosemary, salt and pepper.

Place on an oiled baking sheet and bake until all sides are golden brown and the inside is tender, approximately 40 to 45 minutes. Turn every 15 minutes or so for even browning.

Serves 6

If you've never tried Jerusalem artichokes, or sunchokes as they are sometimes called, then you're in for a treat. They have a wonderful nutty flavor that roasting only enhances.

Savory Bread Pudding with Leeks and Camembert

3 Tbsp.	butter	45 mL
3	leeks, white part only, thinly sliced	3
1	onion, finely sliced	1
1 tsp.	chopped fresh thyme	5 mL
1/4 cup	white wine	60 mL
	salt and freshly ground black pepper to taste	
4	eggs	4
1/4 cup	half-and-half cream	60 mL
2 cups	milk	480 mL
1 tsp.	chopped fresh rosemary	5 mL
1 tsp.	chopped fresh thyme	5 mL
12	1-inch (2.5-cm) slices white bread or egg bread, preferably day-old	12
4 oz.	Camembert cheese, cut into small cubes	113 g
1/4 cup	grated Parmesan cheese	60 mL

Serves 10

It used to be that bread pudding was used to stretch the homemaker's budget, and it was made only for dessert. Like so many things, something old is new again. In this case, bread pudding is served as a savory instead of a sweet. Serve it in lieu of potato or rice as a side dish with meat or fish.

Melt the butter in a large skillet over medium-high heat. Add the leeks, onion and 1 tsp. (5 mL) of the fresh thyme and sauté until lightly golden. Reduce the heat and cook until the vegetables are very soft, approximately 20 minutes. Add the wine and cook until the liquid is reduced by half. Remove from the heat and season with salt and pepper.

Whisk together the eggs, cream, milk, rosemary and remaining 1 tsp. (5 mL) thyme.

Preheat the oven to 325°F (165°C).

Line the bottom of an 8- x 11-inch (20- x 28-cm) buttered baking dish with the bread slices, overlapping them where necessary to cover the bottom completely. Spread 1/2 of the onion mixture over the bread. Sprinkle with 1/3 of the Camembert and Parmesan.

Cover with a second layer of bread, onion and cheeses. Finish with a third layer of bread and sprinkle with the remaining cheeses. Pour the egg mixture over top and press down firmly, making sure the egg mixture covers the bread. Bake in the middle rack of the oven for 1 hour, until golden and puffed.

To serve, cut the pudding into squares.

ONION CONFIT

1/3 cup	*unsalted butter*	80 mL
2 lbs.	*cooking onions, thinly sliced*	900 g
1/2 cup	*sugar*	120 mL
2	*cloves garlic, very finely minced*	2
1/2 cup	*red wine vinegar*	120 mL
2 tsp.	*finely chopped fresh rosemary*	10 mL
1 tsp.	*finely chopped fresh thyme*	5 mL

*Makes approximately
2 cups (480 mL)*

*If you love caramelized
onions, then this is the
dish for you. It can be
used as a side dish for
game or meat dishes or
as a relish for sandwiches
and burgers.*

Melt the butter in a sauté pan over medium heat. When the butter foams, add the onions and sugar and cook until the onions are golden brown, about 10 minutes. Stir frequently.

Add the garlic and continue to cook for 30 seconds. Add the vinegar, rosemary and thyme and continue to cook over medium heat until thickened.

Cool before serving. It can be stored in the refrigerator for up to 1 week.

THREE-GRAIN MUSHROOM PILAF

Serves 6 to 8

This is a great multi-grain recipe which can be used as a vegetarian main course or a terrific side dish. The pilaf reheats beautifully, so make it the day before and enjoy more time with your guests.

2 Tbsp.	olive oil	30 mL
2 cups	sliced mushrooms	480 mL
2 Tbsp.	olive oil	30 mL
4	carrots, cut into small dice	4
1	onion, finely chopped	1
1/3 cup	barley	80 mL
1/3 cup	brown rice	80 mL
1/2 tsp.	coriander seeds, crushed	2.5 mL
1/2 tsp.	cumin seeds, crushed	2.5 mL
2	cloves garlic, finely chopped	2
1/2 cup	dry white wine	120 mL
2 cups	chicken or vegetable stock	480 mL
1/3 cup	slivered almonds, toasted (see page 183)	80 mL
	salt and freshly ground black pepper to taste	
1/3 cup	bulgur wheat	80 mL

Heat 2 Tbsp. (30 mL) of olive oil in a medium saucepan over medium-high heat. Add the mushrooms and sauté until soft and lightly browned. Transfer to a bowl.

Heat the remaining olive oil in the same pan. Add the carrots and onion, and sauté until the onion is soft. Add the barley and the brown rice, stirring until all the grains are coated with oil. Add the coriander seeds, cumin seeds and garlic, and sauté for 30 seconds.

Return the mushrooms to the pan, add the white wine, stock and almonds. Season with salt and pepper.

Bring to a boil, reduce the heat to low, cover and simmer for 30 minutes. Stir in the bulgur and continue to simmer for 10 minutes, or until all the grains are tender.

Remove from the heat and let rest for 5 minutes. Fluff with a fork and transfer to a serving dish.

ROAST ONION CHRYSANTHEMUMS

12	small yellow cooking onions, 2 inches (5 cm) in diameter	12
1/4 cup	chicken stock	60 mL
3 Tbsp.	butter	45 mL
1 tsp.	sugar	5 mL
	salt and freshly ground black pepper to taste	

Serves 6

The method of slicing the onions makes them open up like a flower during cooking. A tool called an onion chrysanthemum cutter makes the task even easier.

Preheat the oven to 425°F (220°C).

Peel the onions. Remove part of the root end of the onion so it will not roll around, but leave enough that it will hold together. Remove a little off the top. Cut down from the top of the onion to within 3/4 inch (2 cm) of the root, being careful not to cut all the way through the onion. Continue to cut in 1/4-inch (.6-cm) intervals, keeping the onion intact. Turn the onion and continue cutting in a cross-hatch pattern. Repeat with all the onions.

Lightly butter an 8- x 15-inch (20- x 38-cm) baking dish. Heat the stock over medium heat. Add the remaining butter.

Place the onions in the dish, giving them lots of room to open while roasting. Sprinkle with sugar, salt and pepper. Pour the stock over the onions. Cover with aluminum foil and bake for 45 minutes, or until the onions are tender. Remove the foil and continue to roast for 30 to 45 minutes, basting occasionally with stock and pan juices until the onions are golden brown.

CARAMELIZED ONION, SHALLOT AND GARLIC TART

For the pastry:

1 1/4 cups	all purpose flour	300 mL
1/2 tsp.	salt	2.5 mL
1/3 cup	chilled butter, cut into pieces	80 mL
2 Tbsp.	chilled vegetable shortening	30 mL
3 Tbsp.	cold water (approximately)	45 mL

Serves 8

Move over Yorkshire pudding. Without a doubt this tart will be a crowd pleaser. If there's anyone who claims not to like onions, let them try these onions and shallots sautéed with garlic until very sweet and served up as a tart. Slice the tart into small wedges to serve it.

Combine the flour and salt. Add the butter and shortening and combine in a food processor or by hand with a pastry blender until the mixture resembles coarse meal. Add the water 1 Tbsp. (15 mL) at a time until the dough comes together into a ball. Flatten into a disk, wrap in plastic wrap and refrigerate for at least 30 minutes, or up to 2 days.

Preheat the oven to 400°F (200°C).

Roll the dough into a 12-inch (30-cm) round on a lightly floured surface. Transfer to a 9-inch (23-cm) tart pan with a removable bottom. Fit the dough into the pan. Trim the dough, leaving it 1/4 inch (.6 cm) longer than the pan rim. Refrigerate for 15 minutes.

Line the crust with foil or parchment paper and fill with pie weights or dried beans. Bake until the crust is set and beginning to turn golden, about 15 minutes. Remove the pie weights and lining and cool completely on a rack.

To make the tart:

2 Tbsp.	olive oil	30 mL
2	large onions, thinly sliced	2
8 to 10	shallots, thinly sliced	8 to 10
10	large garlic cloves, finely chopped	10
1	large fresh thyme sprig	1
1	bay leaf	1
1/2 cup	dry white wine	120 mL
	salt and freshly ground black pepper to taste	
1 cup	shredded Gruyère cheese, packed	240 mL

Heat the oil in a large skillet over medium heat. Add the onions, shallots, garlic, thyme and bay leaf. Cook slowly until the onions are brown and the mixture is jam-like, about 50 minutes. Stir occasionally and watch that the mixture does not burn. Add the wine; cook and stir until almost all the liquid has evaporated. Season with salt and pepper. Remove from the heat and cool. The filling can be prepared to this point 6 hours ahead.

Remove the bay leaf and thyme sprig from the mixture. Stir in the Gruyère cheese. Place the filling in the crust and bake until browned, approximately 20 minutes. Cool slightly before removing the rim from the tart pan. Serve the tart warm or at room temperature.

Parsnip and Carrot Cakes

1/2 lb.	parsnips, peeled and coarsely grated	225 g
1/4 lb.	carrots, peeled and coarsely grated	113 g
2 Tbsp.	all purpose flour	30 mL
2 Tbsp.	butter, melted	30 mL
	salt and freshly ground black pepper to taste	
	pinch ground mace	
3 Tbsp.	butter	45 mL

Serves 4

These little molded cakes look very elegant accompanying sautéed meats.

Bring a small pot of salted water to a boil. Add the grated parsnips and carrots, and blanch for 1 to 2 minutes, until cooked but still firm. Drain and rinse under cold water. Drain again and allow to cool.

Sift the flour over the cooled vegetables. Mix gently, being careful not to mash them. Add the butter, salt, pepper and mace. Stir briefly.

Line four 4-inch (10-cm) flan tins with parchment paper. Divide the mixture evenly between the tins. Chill in the refrigerator until cool and firm, approximately 45 minutes.

Melt the 3 Tbsp. (45 mL) butter in a medium sauté pan over medium heat. Invert the flans into the sauté pan. Remove the flan tins and the parchment. Gently sauté the flans until golden brown, approximately 5 minutes on each side. Keep them warm until serving time.

Note: If flan rings are not available, use a large jar lid as a guide to make the flans.

Southwest Vegetable Chili
(page 150)

Asparagus and Roasted Red Pepper Flan
(page 164)

Wood-Grilled Pizzas with Pesto
(page 184)

Quesadillas (page 186)

Black Bean with Pineapple Salsa
(page 218)

Butternut Squash Cake with Cream Cheese Frosting (page 200)

Butternut Squash and Cranberry Muffins
(page 199)

Pickled Asparagus Spears
(page 220)

GRATIN OF PARSNIPS AND CELERY ROOT

1	clove garlic, crushed	1
2 Tbsp.	butter	30 mL
2 lbs.	parsnips, peeled and thinly sliced	900 g
1 lb.	celery root, peeled and grated	455 g
	salt and freshly ground black pepper to taste	
	pinch freshly grated nutmeg	
1 1/2 cups	whipping cream	360 mL
1/4 cup	grated Gruyère cheese	60 mL

Serves 6

Two of the more under-rated members of the root vegetable family strut their stuff in this dish. Celery root, also known as celeriac, and parsnips are paired in an elegant gratin.

Preheat the oven to 350°F (175°C).

Rub the bottom and sides of a large gratin dish or a 6-cup (1.5-L) baking dish with the crushed garlic clove. Discard the garlic. Coat the sides and bottom of the gratin dish with the butter.

Combine the parsnips and celery root, salt, pepper and nutmeg in a large mixing bowl. Place the vegetables in the prepared gratin dish and gently pat them down. Pour the cream over the top. Bake for 30 minutes.

Remove from the oven and discard any brown crust that may have formed. Sprinkle with the grated cheese and return to the oven for another 30 minutes, or until golden brown on top and tender inside.

Green Pea, Cauliflower and Tomato Curry

Serves 4 to 6

This authentic Indian curry is very quick and easy to prepare. Serve it with grilled meat and a bowl of steaming rice.

1	small head cauliflower, separated into florets	1
1 1/2 cups	fresh or frozen peas	360 mL
4 Tbsp.	butter	60 mL
1	onion, finely chopped	1
1	clove garlic, finely chopped	1
1 tsp.	grated fresh ginger	5 mL
1 Tbsp.	ground coriander	15 mL
1 Tbsp.	curry paste or powder	15 mL
3/4 tsp.	turmeric	4 mL
2	whole cardamom pods, slightly crushed	2
3/4 cup	coconut milk	180 mL
2	tomatoes, cut into 8 wedges each	2
2 tsp.	sugar	10 mL
	salt and freshly ground black pepper to taste	

Bring a large pot of salted water to a boil. Add the cauliflower and fresh peas. Cook until tender and drain. (If using frozen peas, just thaw them and add them to the cooked cauliflower.)

Melt the butter in a medium skillet over medium-high heat. Add the onion and sauté until translucent. Add the garlic and ginger, and sauté for 30 seconds. Add the coriander, curry, turmeric and cardamom. Sauté for 30 seconds. Add the coconut milk, tomatoes and sugar. Simmer for 5 minutes.

Add the cauliflower and peas, and continue to cook until heated through, 4 to 5 minutes. Season with salt and pepper. Serve hot.

POTATOES DAUPHINOISE

1	clove garlic, halved	1
2 Tbsp.	butter	30 mL
1 1/2 cups	milk, more or less	360 mL
2 lbs.	potatoes, peeled and very thinly sliced	900 g
	salt and freshly ground black pepper	
	to taste	
	pinch freshly grated nutmeg	
1/2 cup	whipping cream	120 mL
1/4 cup	grated Gruyère cheese	60 mL

Serves 6

The ultimate dress-me-up-and-show-me-off potatoes.

Preheat the oven to 350°F (175°C).

Rub a shallow 10-inch (25-cm) gratin dish with the cut sides of the garlic. Coat the sides and bottom of the dish with the butter.

Place the dish over medium-high heat and add 3/4 cup (180 mL) of the milk. As the milk is heating, quickly place the potato slices in concentric circles, sprinkling with salt and pepper now and then. When all the potato is used up, add more milk if necessary, enough to barely cover the potatoes. Grate a little nutmeg on top.

Line a baking sheet with aluminum foil and place the gratin dish on the sheet to protect the oven from spills. Bake until the milk is all absorbed, approximately 45 minutes.

Remove the gratin from the oven and remove the brown skin that has formed over the potatoes. Pour the cream over the potatoes. Sprinkle with a little salt and pepper and the grated cheese. Return to the oven and continue to bake until the cream is thick and the cheese is bubbly and brown on top. Let stand 5 minutes before serving.

NEW POTATOES WITH MUSTARD-DILL MAYONNAISE

Serves 4

This versatile dish can be served hot as a side dish, or at room temperature on a buffet table—and it makes great picnic fare.

1 lb.	*very small new potatoes, scrubbed*	*455 g*
1/4 cup	*store-bought or Homemade Mayonnaise (page 214)*	*60 mL*
2 Tbsp.	*olive oil*	*30 mL*
1 Tbsp.	*white wine vinegar*	*15 mL*
1 Tbsp.	*Russian-style mustard*	*15 mL*
2 Tbsp.	*finely chopped fresh dill*	*30 mL*
2 Tbsp.	*green onion, finely chopped*	*30 mL*
	salt and freshly ground black pepper to taste	
1 Tbsp.	*fresh chopped dill or chives (optional)*	*15 mL*

Place the potatoes in a saucepan, cover with salted water and bring to a boil. Reduce the heat and simmer until the potatoes are just tender. Drain and return the potatoes to the pan to dry a little.

While the potatoes are cooking, make the dressing.

Combine the mayonnaise, oil, vinegar, mustard and dill in a small bowl and whisk until well combined. Pour the dressing over the warm potatoes, add the green onions and toss well. Season with salt and pepper.

Garnish with chopped fresh dill or chopped fresh chives, if desired.

POTATO LATKES

2 lbs.	potatoes	900 g
2	large eggs	2
2 Tbsp.	all purpose flour	30 mL
1 tsp.	salt	5 mL
1–3 tsp.	grated onion	5–15 mL
3 Tbsp.	vegetable oil	45 mL
1 cup	sour cream and/or applesauce (optional)	240 mL

Makes 16

Peel and finely grate the potatoes, immersing them immediately in cold water. Drain, wrap in a tea towel and squeeze as much moisture from the potatoes as you can.

Beat the eggs well in a large bowl. Add the grated potato, flour and salt, mixing well. Mix in the grated onion.

Heat the oil in a heavy skillet over medium-high heat. Drop the batter by spoonfuls onto the hot skillet, flattening it out to make 3- to 5-inch (7- to 12-cm) circles. When the cake has browned, turn it over and brown the other side. Serve warm.

Serve bowls of sour cream and applesauce as an accompaniment to the latkes.

These little potato pancakes, sometimes described as Jewish soul food, can be served as a side dish or as a base for canapés if made smaller. This is probably one of the most famous dishes of Jewish cookery and one of the easiest to make.

Mashed Potatoes with Horseradish

6	medium potatoes, peeled and cut into cubes	6
1/2 cup	warm milk	120 mL
1/4 cup	butter	60 mL
2 Tbsp.	prepared horseradish	30 mL
	salt and freshly ground black pepper to taste	

Serves 4 to 6

These are unbelievably good—the perfect accompaniment to a medium-rare roast of beef.

Place the potatoes in a large saucepan and cover with salted water. Bring to a boil and cook until the potatoes are tender, approximately 15 minutes. Drain.

Return the potatoes to the same pot, and leave them to dry for a minute. Mash the potatoes with a masher or use a food mill, being sure to keep them warm. Add the warm milk, butter, horseradish, salt and pepper. Mix well.

Keep the potatoes warm until serving time. (When potatoes are allowed to get cold the starch clumps, causing lumpy potatoes.)

COCONUT RICE

1/4 cup	butter	60 mL
1	onion, finely chopped	1
2 cups	long-grain rice	480 mL
2 cups	coconut milk	480 mL
1/2 cup	water	120 mL
1/2 tsp.	salt	2.5 mL
1/2 cup	unsweetened coconut	120 mL
1 tsp.	butter	5 mL

Serves 6

Melt the butter in a medium saucepan with a tight-fitting lid over medium-high heat. When it foams, add the onion and sauté until it's translucent. Add the rice and sauté, stirring constantly until all the grains are coated with butter. Add the coconut milk, water and salt. Bring to a boil, reduce the heat to low, cover and simmer for 18 minutes.

Meanwhile, toast the coconut in a sauté pan over medium heat. Stir constantly. Remove from the pan when it just starts to brown. Remove the rice from the heat and let it sit for 5 minutes. Remove the lid. Add the butter to the rice and fluff it with a fork. Transfer to a serving dish and sprinkle the toasted coconut on top.

This is an old standby in our house. If we're having curry we're having coconut rice. Use any leftovers in rice pudding.

Thai Rice with Basil and Mint

Serves 6

Stir-fry a little shrimp or chicken to go along with this dish and you've got a great meal.

1 1/2 cups	jasmine or basmati rice	360 mL
1 1/2 cups	chicken stock	360 mL
1 cup	coconut milk	240 mL
2 Tbsp.	sherry	30 mL
2 Tbsp.	fish sauce	30 mL
2 tsp.	grated orange zest	10 mL
1 tsp.	grated lime zest	5 mL
2 Tbsp.	vegetable oil	30 mL
2	cloves garlic, minced	2
1 Tbsp.	grated lemon zest	15 mL
1/2 cup	currants, washed and dried	120 mL
1 Tbsp.	butter	15 mL
1	small sweet red pepper, finely diced	1
3	green onions, cut on the diagonal into 1/4-inch (.6-cm) pieces	3
1/4 cup	chopped fresh basil	60 mL
1/4 cup	chopped fresh mint	60 mL
1/4 cup	chopped unsalted peanuts	60 mL

Rinse the rice under cold water until the water runs clear. Drain and set aside.

Mix the stock, coconut milk, sherry, fish sauce and orange and lime zest together in a bowl. Set aside.

Heat the oil in a medium saucepan with a tight-fitting lid over medium-high heat. Add the garlic and lemon zest and sauté for 30 seconds. Add the rice and stir until all the grains of rice are covered with oil. Stir the currants into the rice mixture. Add the stock mixture and bring to a boil. Reduce the heat to low, cover and simmer for 20 minutes, or until the rice is tender and the stock is absorbed.

Remove from the heat and let rest for 5 minutes. Remove the lid, add the butter and fluff the rice with a fork.

Toss the rice with the red pepper, green onion, basil and mint. Sprinkle with chopped peanuts just before serving.

WILD RICE WITH PINE NUTS AND SUN-DRIED CRANBERRIES

4 cups	water	1 L
1 tsp.	salt	5 mL
1 cup	wild rice	240 mL
1/2 cup	sun-dried cranberries	120 mL
1/4 cup	cognac or brandy	60 mL
2 Tbsp.	butter	30 mL
1	shallot, finely chopped	1
1/4 cup	pine nuts	60 mL
	salt and freshly ground black pepper to taste	

Serves 6

This delicious side dish can also be used as a stuffing for poultry or pork. Cool the mixture before stuffing the meat.

Bring the water to a boil in a medium saucepan. Add the salt and wild rice. Cover, reduce the heat and simmer until the rice is tender, approximately 30 minutes. Drain the rice and set aside.

Soak the cranberries in the cognac or brandy for 10 minutes. Drain, reserving the cognac.

Melt the butter in a medium sauté pan over medium heat. Sauté the shallot for 1 minute. Add the cranberries and pine nuts. Stir in the rice and reserved cognac. Season with salt and pepper to taste and serve warm.

RICE VERDE

1/2 cup	*fresh coriander, tightly packed*	120 mL
1 cup	*fresh spinach, tightly packed*	240 mL
1 1/4 cups	*chicken stock*	300 mL
1 1/4 cups	*milk*	300 mL
1 tsp.	*salt*	5 mL
1/4 cup	*butter*	60 mL
1	*onion, finely chopped*	1
1	*clove garlic, finely chopped*	1
1 1/2 cups	*long-grain rice*	360 mL
1 tsp.	*butter*	5 mL

Serves 4 to 6

The perfect accompaniment to Tex-Mex or Spanish dishes.

Purée the coriander, spinach and stock in a blender. Add the milk and salt and mix until combined.

Melt the 1/4 cup (60 mL) butter in a medium saucepan with a tight-fitting lid over medium-high heat. When it foams, add the onion and sauté until it's translucent. Add the garlic and sauté 30 seconds. Add the rice and stir until all the grains are coated. Add the purée mixture, and bring to a boil. Reduce the heat to low, cover and simmer for 18 minutes.

Remove the rice from the heat and let it sit, covered, for 5 minutes. Add the 1 tsp. (5 mL) of butter to the rice and fluff it with a fork. Serve immediately.

BUTTERED SALSIFY

3 cups	water	720 mL
1 Tbsp.	flour	15 mL
2 tsp.	lemon juice	10 mL
1/2 tsp.	salt	2.5 mL
2 cups	salsify, peeled and sliced into 1-inch (2.5-cm) lengths	80 mL
3 Tbsp.	butter	45 mL
	fresh grating of nutmeg	
	salt and freshly ground black pepper to taste	

Bring the water to a boil over high heat. Add the flour, lemon juice and salt. Mix well, add the salsify and cook until just barely tender, 7 to 10 minutes. Drain.

Melt the butter in a medium sauté pan over medium-high heat. Add the salsify and sauté for 1 or 2 minutes. Season with nutmeg, salt and pepper and transfer to a serving dish.

Serves 4

If you've never tried this black-skinned, carrot-shaped vegetable you're in for a treat. Salsify tastes like hot buttered popcorn. To prevent discoloring, it's boiled with a little flour and lemon juice.

CREAMED SPINACH

2 lbs.	*fresh spinach, washed and coarse stems removed*	900 g
2 Tbsp.	*unsalted butter*	30 mL
2	*shallots, finely chopped*	2
1/2 cup	*whipping cream*	120 mL
	pinch freshly grated nutmeg	
	salt and freshly ground black pepper to taste	

Serves 4

This recipe can easily be doubled for a larger crowd. Serve it as a side dish with roasted meats or poultry.

Bring 1 inch (2.5 cm) of lightly salted water to a boil in a large saucepan. Add the spinach and cook until it's completely wilted. Drain in a colander, pressing it with a spoon to remove as much water as possible.

Melt the butter in a medium sauté pan over medium-high heat. Add the shallot and sauté until it's translucent. Add the spinach and sauté for 30 seconds. Add the cream and nutmeg. Season with salt and pepper. Continue to cook until most of the cream is absorbed.

Spinach and Plum Tomatoes with Gruyère Cheese

1/4 cup	vegetable oil	60 mL
1	medium onion, finely chopped	1
2	cloves garlic, finely chopped	2
2 cups	coarsely chopped plum tomatoes	480 mL
8 cups	fresh spinach, washed and coarse stems removed	2 L
	salt and freshly ground black pepper to taste	
1 1/2 cups	grated Gruyère cheese	360 mL

Serves 4 to 6

If you're looking for a quick and delicious side dish, this is it.

Heat the oil in a large saucepan over medium heat. Add the onion and sauté until it's translucent. Add the garlic and sauté for 30 seconds. Add the tomatoes and simmer until all the liquid has evaporated. Stir in the spinach, cooking until it has wilted. Season with salt and pepper. Toss with 1 cup (240 mL) of the cheese. Remove from the heat.

Transfer the mixture to an oiled ovenproof dish. Sprinkle the remainder of the cheese on top. Place under the broiler until the cheese has lightly browned.

ROASTED BUTTERNUT SQUASH CASSEROLE

Serves 12

This casserole can be made ahead, reheated and served the next day, making it a great dish for potluck suppers.

2 1/2 lbs.	butternut squash, cut in half and seeds discarded	1.1 kg
1/4 cup	butter	60 mL
4 cups	sliced onions	1 L
	pinch sugar	
2	cloves garlic, finely chopped	2
3	egg yolks	3
3	eggs	3
1 1/2 cups	whipping cream	360 mL
1 Tbsp.	chopped fresh thyme	15 mL
	freshly grated nutmeg	
	salt and freshly ground black pepper to taste	

Preheat the oven to 400°F (200°C).

Prick the squash with a fork and place cut side down on a lightly oiled baking sheet. Bake until tender, approximately 45 to 50 minutes. Let cool, remove the skin and cut the squash into 1-inch (2.5-cm) cubes. Set aside.

Melt the butter in a medium sauté pan over medium-high heat. Add the onions and pinch of sugar. Sauté, stirring frequently, until the onions are translucent. Reduce the heat to low and continue cooking, stirring frequently, until the onions are golden brown, approximately 20 minutes. Add the garlic and cook 1 minute longer. Remove from the heat.

Preheat the oven to 350°F (175°C).

Lightly butter a 9- x 13-inch (23- x 33-cm) baking dish. Toss the squash and onions together, and place them in the baking dish.

In a medium bowl whisk the egg yolks, eggs, cream, thyme, nutmeg, salt and pepper. Pour the mixture over the squash. Cover with aluminum foil and bake for 45 minutes. Remove the foil and bake 5 minutes longer, or until the custard is set.

Baked Trio of Squash

1	small turban squash, cut into 1-inch-thick (2.5-cm) slices and seeds discarded	1
1	small dumpling or butternut squash, cut into 1-inch-thick (2.5-cm) slices and seeds discarded	1
1	acorn squash, cut into 1-inch-thick (2.5-cm) slices and seeds discarded	1
2 Tbsp.	butter, melted	30 mL
	salt and freshly ground black pepper to taste	
1/3 cup	butter	80 mL
1/3 cup	honey	80 mL
1/2 tsp.	ground ginger	2.5 mL
	pinch ground cinnamon	

Serves 8 to 10

Leaving the skins on the squash gives this dish its appeal. Use any winter squash that are approximately the same size. This recipe goes especially well with roast beef or roast poultry.

Preheat the oven to 350°F (175°C).

Lightly brush the squash slices with the 2 Tbsp. (30 mL) of melted butter. Place on a parchment-lined baking sheet. Season with salt and pepper. Bake until the squash is tender, approximately 35 to 40 minutes, turning once and basting with a little more butter.

Melt the 1/3 cup (80 mL) of butter in a small saucepan over medium heat and add the honey, ginger and cinnamon. Heat until well combined.

Remove the squash from the oven and brush with the honey-butter mixture. Arrange on a serving platter and serve immediately.

BAKED SWEET POTATO WITH LEEKS AND APPLES

Serves 6

Sweet potato, leeks and apple—reminiscent of fall and harvest time—combine perfectly for a Thanksgiving feast.

1 1/2 cups	*whipping cream*	*360 mL*
2	*leeks, white part only, thinly sliced*	*2*
1 Tbsp.	*chopped fresh thyme*	*15 mL*
1/8 tsp.	*ground cinnamon*	*.5 mL*
	pinch freshly grated nutmeg	
	salt and freshly ground black pepper to taste	
2	*medium sweet potatoes, peeled and thinly sliced*	*2*
4	*Golden Delicious apples, peeled, cored and thinly sliced*	*4*
1 cup	*grated Cheddar cheese*	*240 mL*

Place the cream and leeks in a medium saucepan and bring to a boil. Add the thyme, cinnamon, nutmeg, salt and pepper. Simmer until the leeks are soft, approximately 10 minutes.

Preheat the oven to 350°F (175°C).

Lightly butter a 1-quart (1-L) baking dish. Place 1/3 of the potatoes in the bottom, ladle a little of the leek mixture on top, and cover with 1/2 the apples. Repeat the layers, ending with a layer of potatoes and the last of the leek mixture on the top.

Bake covered for 45 minutes. Uncover, sprinkle with the cheese and continue to bake for 30 minutes, or until the potatoes are tender and brown on top. Let rest a few minutes before serving.

SWISS CHARD WITH RAISINS AND PINE NUTS

2 lbs.	Swiss chard	900 g
2 Tbsp.	raisins	30 mL
1 Tbsp.	olive oil	15 mL
1	clove garlic, finely chopped	1
	pinch salt	
2 tsp.	red wine vinegar	10 mL
2 Tbsp.	pine nuts, toasted (see page 183)	30 mL

Serves 6

A unique way to enjoy your greens, with the sweetness of raisins and the nutty flavor of pine nuts.

Thoroughly wash the Swiss chard. Remove the midribs from the leaves and set the leaves aside. Chop the ribs into small pieces.

Steam the chard ribs until they are just beginning to be tender, about 3 to 5 minutes. Add the leaves and the raisins and cook until the leaves are wilted and the stems are completely tender, another 3 minutes.

Heat the olive oil in a frying pan over medium heat. Add the garlic and sauté 30 seconds. Add the chard, raisins and salt and stir just until heated through. Remove the pan from the heat and stir in the vinegar. Place in a serving dish, sprinkle with pine nuts, and serve immediately.

TOMATOES STUFFED WITH SPINACH AND PINE NUTS

Serves 6

Stuffed tomatoes make an elegant addition to any dinner and look especially tempting on a hot buffet table.

6	large tomatoes	6
1/4 cup	butter	60 mL
1/2 cup	finely chopped onion	120 mL
2	10-oz. (285-g) packages frozen chopped spinach, cooked, drained and gently squeezed	2
1/2 tsp.	dried thyme	2.5 mL
1/4 cup	grated Parmesan cheese	60 mL
1/4 cup	soft bread crumbs	60 mL
1/2 cup	pine nuts, toasted (see page 183)	120 mL
2	eggs, lightly beaten	2
	salt and freshly ground black pepper to taste	
1/4 cup	grated Parmesan cheese	60 mL

Bring a pot of water to a boil. Make a small X in the bottom of each tomato with a sharp knife. Drop the tomatoes into the boiling water. After 1 minute, remove them and plunge them into ice water for a few minutes. The skins should peel off easily.

Remove the tops from the tomatoes and carefully scoop out the seeds. Invert the tomatoes on paper towel and let them drain while mixing the filling.

Preheat the oven to 350°F (175°C).

Melt the butter in a medium sauté pan over medium heat. Add the onion and cook until it's translucent. Add the spinach, thyme, 1/4 cup (60 mL) of Parmesan cheese, bread crumbs and pine nuts. Mix well. Add the beaten eggs to the mixture and continue to stir until it's well combined and the eggs are just cooked through. Season with salt and pepper.

Place the tomatoes in a shallow baking dish. Stuff the tomatoes with the spinach filling. Sprinkle with the remaining 1/4 cup (60 mL) of Parmesan cheese.

Bake for 10 minutes, or until golden brown on top and heated through. Serve immediately.

Sweet Potato Batons

2 lbs.	sweet potatoes	900 g
	vegetable oil	
	salt to taste	

Peel the sweet potatoes and cut them into sticks, 1/4 to 1/2 inch (.6 to 1.2 cm) thick.

Serves 4 to 6

Heat the oil in a medium saucepan over medium-high heat to 325°F (165°C). Working in batches, deep-fry the potatoes until light golden brown on the edges. Remove the potatoes with a slotted spoon and drain on paper towel.

Using sweet potatoes instead of regular potatoes gives a new lift to an old idea. Deep-frying them a second time gives these fries (and traditional potatoes) a wonderful "chip shop" flavor.

Increase the oil temperature to 375°F (190°C). Return the potatoes to the oil and cook until they're a deep golden-brown color. Remove with a slotted spoon and drain on paper towel. Sprinkle with salt. Serve immediately with Homemade Mayonnaise (page 214).

Oven-Baked Turnip and Carrot

3 cups	turnip, peeled and cubed	720 mL
3 cups	carrot, peeled and sliced	720 mL
2 Tbsp.	butter	30 mL
2	eggs, lightly beaten	2
3 Tbsp.	flour	45 mL
1 tsp.	brown sugar	5 mL
1 tsp.	baking powder	5 mL
1/2 tsp.	salt	2.5 mL
	freshly ground black pepper to taste	
	pinch freshly grated nutmeg	

Serves 6

The carrot helps sweeten the turnip in this dish. Try it on people who hate turnip—it may change their minds. This is one of those recipes that can be made ahead and served the next day.

Bring two separate saucepans of salted water to a boil. Place the turnip in one and the carrots in the other. Cook until tender. Drain and mash the carrots and turnips together. Add the butter and eggs and stir until well combined. Set aside.

Preheat the oven to 375°F (190°C).

Lightly butter a 1- to 2-quart (1- to 2-L) casserole dish.

In a separate bowl mix the flour, sugar, baking powder, salt, pepper and nutmeg. Combine with the turnip and carrot mixture, mixing well. Transfer to the prepared casserole dish. Bake for 25 to 30 minutes, or until it's beginning to brown on top.

Oven-Roasted Root Vegetables

2	medium sweet potatoes, peeled and cut into 1/4-inch (.6-cm) rounds	2
6	parsnips, peeled, cut in half and then cut lengthwise	6
6	carrots, peeled, cut in half and then cut lengthwise	6
6	shallots, peeled and cut in half	6
1	red onion, peeled and cut into 6 wedges	1
1	large head garlic, cloves separated and peeled	1
2 Tbsp.	chopped fresh rosemary	30 mL
2 Tbsp.	chopped fresh thyme	30 mL
4 Tbsp.	vegetable oil	60 mL
	salt and freshly ground black pepper to taste	

Serves 6

So simple but possibly the best way to enjoy root vegetables. The slow roasting brings out all the sweetness of the vegetables. Perfect with roasted meat or poultry.

Preheat the oven to 375°F (190°C).

Place all the vegetables, including the garlic, in a large bowl. Add the rosemary and thyme, sprinkle with the vegetable oil and toss until well coated.

Place the vegetables on a lightly oiled baking sheet. Season with salt and pepper. Roast, turning frequently, until the vegetables are golden brown and tender, about 45 to 60 minutes.

Transfer to a warmed serving platter, season again with salt and pepper and serve immediately.

THAI STIR-FRIED VEGETABLES

Serves 4 to 6

Serve this as a spicy side dish, or spoon it over rice for a vegetarian entrée.

1/4 cup	vegetable oil	60 mL
1 tsp.	grated fresh ginger	5 mL
3	cloves garlic, crushed	3
1	small red onion, cut into 8 wedges	1
1	large carrot, peeled and very thinly sliced on the diagonal	1
1 cup	cauliflower, cut into small pieces	240 mL
1 cup	zucchini, thinly sliced on the diagonal	240 mL
1 cup	trimmed snow peas	240 mL
1/2 tsp.	salt	2.5 mL
1 Tbsp.	sugar	15 mL
1 Tbsp.	soy sauce	15 mL
2 Tbsp.	Thai fish sauce (nam pla), available at Asian grocery stores	30 mL
2 Tbsp.	lime juice	30 mL
1 tsp.	crushed chili peppers	5 mL
2	green onions, sliced on the diagonal	2
2 Tbsp.	chopped fresh coriander	30 mL

Heat the oil in a wok over medium-high heat. Add the ginger, garlic and onion and stir-fry until the onion is beginning to get soft. Add the carrot and cauliflower and stir-fry for 2 minutes longer. Add the zucchini, snow peas, salt, sugar, soy sauce, fish sauce, lime juice and chili peppers. Continue to stir-fry until the vegetables are barely tender. Do not overcook.

Place in a serving bowl, sprinkle the green onions and coriander on top and serve immediately.

MAIN DISHES

A shift in North American cuisine has seen
vegetables attain greater status in the kitchen.
Their different colors, tastes and textures—
not to mention heathy qualities—have led us all
to reconsider our choices at the dinner table.

The following main course dishes present the
once-lowly vegetable in a brighter light, as
something to be celebrated rather than endured.

These recipes include stews, casseroles, egg dishes,
vegetable and grain dishes such as pilafs and
risottos, savory tarts and pies, and finally
pasta, pizza and quesadillas.

COUNTRY POTATO FRITTATA

Serves 6

Sort of an omelette with-out the fold, frittata hails from Italy. A wonderful brunch dish, it can be served hot directly from the pan and makes terrific picnic fare when cut into squares and served cold.

2	*medium potatoes, scrubbed*	2
10	*large eggs*	10
2 Tbsp.	*chopped fresh rosemary*	30 mL
2 Tbsp.	*chopped fresh thyme*	30 mL
3 Tbsp.	*vegetable oil*	45 mL
1	*small onion, thinly sliced*	1
1/2 cup	*ham, cut into 1-inch (2.5-cm) cubes*	120 mL
1/2 cup	*grated Gruyère cheese*	120 mL
	salt and freshly ground black pepper to taste	

Place the potatoes in a saucepan with salted water. Boil just until tender, drain and set aside to cool.

Beat the eggs in a large mixing bowl until well mixed. Stir in the rosemary and thyme.

Preheat the oven to 400°F (200°C).

Pour the oil into a 2-quart (2-L) baking dish or a 10-inch (25-cm) cast-iron frying pan, coating it evenly, including the bottom and sides. Scatter the onion on the bottom of the dish. Place in a hot oven for 10 minutes or until the onion is softened but not browned. Remove the dish from the oven and reduce the temperature to 375°F (190°C).

Arrange the ham in a single layer on top of the onions. Slice the potatoes thinly and arrange on top of the ham. Pour in the egg mixture and sprinkle the top with the cheese. Season with salt and pepper.

Return the dish to the oven and bake for 25 to 30 minutes, or until the eggs are set and the frittata is puffed and lightly browned on top. Serve hot, warm or cold.

Note: For a vegetarian dish, replace the ham with a 10-oz. (285-g) package of frozen spinach, thawed and chopped.

OLD-FASHIONED BAKED BEANS

1 lb.	dried pinto beans	455 g
6 cups	cold water	1.5 L
1/2 lb.	bacon, chopped	225 g
1	large onion, chopped	1
2	cloves garlic, minced	2
1 cup	ketchup or chili sauce	240 mL
2/3 cup	maple syrup	180 mL
1 Tbsp.	molasses	15 mL
1 Tbsp.	Dijon mustard	15 mL
1/2 tsp.	crushed hot pepper flakes	2.5 mL
1/2 tsp.	Worcestershire sauce	2.5 mL
	pinch baking soda	

Serves 8 to 12

Baked beans have been around for generations, and many families have developed their own special adaptations that they claim to be the best. Bacon gives this dish its smoky appeal, and using pinto beans instead of the conventional navy beans gives it a richer look. This rustic main course can also be used as a side dish.

Wash and sort the beans, discarding any blemished ones. Place the beans and 6 cups (1.5 L) of cold water in a large saucepan and let sit overnight.

The next day, bring it to a boil and boil for 2 minutes. Drain and add 5 cups (1.2 L) of water. Bring it to a boil, reduce the heat, cover and simmer for 1 1/2 hours, or until the beans are tender. Drain off the liquid, reserving 1 1/2 cups (360 mL).

Sauté the bacon in a skillet over medium heat until it begins to brown. Add the onion and garlic and continue to cook until the onion is just translucent and the bacon is beginning to crisp.

Preheat the oven to 250°F (120°C).

Place the beans, onion, garlic and bacon along with all the pan drippings in a large casserole with a tight-fitting lid. Add the reserved 1 1/2 cups (360 mL) bean liquid, ketchup or chili sauce, maple syrup, molasses, mustard, hot pepper flakes, Worcestershire sauce and baking soda. Mix well. Cover and bake for 2 hours, stirring occasionally. Uncover and bake for 30 minutes longer, or until the beans are very tender and slightly thickened.

Note: For a vegetarian version, omit the bacon and use 2 Tbsp. (30 mL) vegetable oil to sauté the garlic and onion.

SPINACH SOUFFLÉ WITH CHEDDAR AND WALNUTS

Serves 4

This classic dish has stood the test of time. I've put a little different spin on it by adding walnuts and a Cheddar cheese sauce.

1 Tbsp.	unsalted butter, softened	15 mL
1/4 cup	walnuts, toasted (see page 183) and very finely ground	60 mL
1 Tbsp.	unsalted butter	15 mL
1	small onion, finely chopped	1
1 cup	steamed fresh spinach, drained and finely chopped, or 1 10-oz. (285-g) pkg. frozen spinach, thawed, drained and finely chopped	240 mL
	pinch cayenne pepper	
	pinch freshly grated nutmeg	
1/4 tsp.	salt	1.2 mL
6 Tbsp.	unsalted butter	90 mL
6 Tbsp.	all purpose flour	90 mL
2 cups	milk	480 mL
1/4 tsp.	salt	1.2 mL
4	egg yolks	4
1/2 cup	grated Cheddar cheese	120 mL
6	egg whites	6
1 cup	grated Cheddar cheese	240 mL

Preheat the oven to 375°F (190°C).

With 1 Tbsp. (15 mL) of butter, grease a 6- to 8-cup (1.5- to 2-L) soufflé dish. Sprinkle the bottom and sides with the ground walnuts. Reserve any of the nuts that do not stick.

Melt 1 Tbsp. (15 mL) of butter in a medium saucepan over medium heat. Add the onion and sauté until it becomes translucent. Add the spinach, cayenne pepper, nutmeg, 1/4 tsp. (1.2 mL) salt and any reserved walnuts. Set aside.

Melt the remaining 6 Tbsp. (90 mL) butter in a medium saucepan over medium heat. Add the flour and stir until well blended. Slowly add the milk, cooking and stirring as the sauce thickens. Add the remaining 1/4 tsp. (1.2 mL) of salt. When all the milk is incorporated and has thickened, pour half the sauce into a bowl and set aside.

To the sauce remaining in the pot, add the egg yolks one at a time, stirring constantly. Return the mixture to the heat and continue to cook for 2 minutes longer over low heat. Fold in the spinach mixture. Remove from the heat and cool slightly. Add 1/2 cup (120 mL) of Cheddar cheese.

Beat the egg whites until they are stiff but not dry. Fold the egg whites into the spinach mixture. Pour into the prepared dish and bake until the soufflé is firm and golden brown, approximately 30 to 40 minutes.

While the soufflé is baking, heat the reserved sauce very slowly over low heat. Add the remaining 1 cup (240 mL) of grated cheese. Stir until it has melted and the sauce is smooth.

Serve the soufflé as soon as it comes out of the oven, with the cheese sauce alongside.

BOMBAY VEGETABLE CURRY

Serves 6

The evocative aroma of curry cooking on the stove—this is definitely not the way my mom cooked vegetables. Serve this dish over steamed rice for a light and satisfying dinner.

1/3 cup	vegetable oil	80 mL
1 tsp.	cumin seeds	5 mL
1 tsp.	coriander seeds	5 mL
1 tsp.	fenugreek seeds	5 mL
2 tsp.	finely chopped fresh ginger	10 mL
1	clove garlic, finely chopped	1
1/4 tsp.	cayenne pepper	1.2 mL
1 tsp.	ground turmeric	5 mL
1/3 cup	flour	80 mL
2 Tbsp.	tomato paste	30 mL
10	pearl onions, skins removed	10
2	sweet green peppers, chopped into large pieces (or 1 red and 1 green pepper)	2
1	carrot, cut into 1-inch (2.5-cm) pieces	1
1	baby eggplant, cut into 1-inch (2.5-cm) slices	1
8 cups	vegetable or chicken stock, heated	2 L
1	medium sweet potato, cut into large cubes	1
3	medium potatoes, cut into quarters	3
10	pods okra, sliced into 1-inch (2.5-cm) pieces	10
1 cup	green or yellow beans, trimmed	240 mL
	pinch sugar	
	salt and freshly ground black pepper to taste	
2 Tbsp.	freshly chopped coriander (optional)	30 mL

Heat the oil in a large saucepan over medium heat. Add the cumin, coriander and fenugreek, and sauté for 30 seconds. Add the ginger, garlic, cayenne pepper and turmeric, and sauté for 30 seconds. Add the flour and stir until all the ingredients become a paste. Add the tomato paste and sauté for 30 seconds.

Add the pearl onions, green pepper and carrot, and sauté for 3 minutes, stirring frequently and scraping up any brown bits that form on the bottom of the pan. Add the eggplant and sauté for 1 minute more.

Add the stock, sweet potatoes and potatoes. Bring the mixture to a boil, reduce the heat and simmer for 5 minutes. Add the okra and green or yellow beans. Simmer until all the vegetables are tender. Add a pinch of sugar, and season with salt and pepper.

Before serving, sprinkle with fresh coriander, if desired.

SOUTHWEST VEGETABLE CHILI

**Serves 6
generously**

*Don't tell anyone this is
all vegetable, they'll never
know the difference. Serve
it with tortilla chips and
garnish it with grated
Cheddar cheese for a
casual dinner. It's even
better if refrigerated
overnight and reheated
the next day. You can
turn up the heat by
adding more jalapeño
pepper or chili powder.*

2	14-oz. (398-mL) cans red kidney beans, drained	2
2	28-oz. (796-mL) cans plum tomatoes	2
1/4 cup	vegetable oil	60 mL
1	medium onion, coarsely chopped	1
2	carrots, cut into small dice	2
1	celery stalk, cut into small dice	1
1	sweet red pepper, roughly diced	1
1	sweet green pepper, roughly diced	1
1	jalapeño pepper, seeded and finely diced	1
2	cloves garlic, finely chopped	2
1/4 cup	tomato paste	60 mL
1	12-oz. (340-mL) bottle beer	1
3 Tbsp.	chili powder	45 mL
1 tsp.	cumin	5 mL
1 tsp.	oregano	5 mL
2 Tbsp.	brown sugar	30 mL
	salt and freshly ground black pepper to taste	

Pulse 1 can of kidney beans 3 or 4 times in a food processor.
Set aside.

Process 1 can of tomatoes in a food processor until puréed.
Set aside.

Heat the oil in a large saucepan over medium-high heat. Add the
onion, carrot and celery, and sauté for 3 minutes, or until the
onion becomes translucent. Add the red, green and jalapeño
peppers, and sauté for 2 minutes. Add the garlic and sauté for
30 seconds. Add the tomato paste and cook for 1 minute, or until
the paste begins to darken. Add the chopped kidney beans and
sauté for 1 minute longer. Add the remaining can of kidney beans,
the puréed tomatoes, the remaining can of tomatoes, beer, chili
powder, cumin, oregano and brown sugar.

Bring to a boil, stirring often to prevent sticking. Reduce the heat to low and simmer for 1 hour, stirring occasionally.

Just before serving, adjust the seasoning with salt and pepper.

Mayan Stew

1/4 cup	vegetable oil	60 mL
1	medium onion, finely chopped	1
2	cloves garlic, finely chopped	2
1 Tbsp.	sweet paprika	15 mL
2 tsp.	chili powder	10 mL
1 tsp.	each ground cumin and coriander	5 mL
1 tsp.	dried oregano	5 mL
1/2 tsp.	ground cinnamon	2.5 mL
1/4 tsp.	freshly grated nutmeg	1.2 mL
	pinch ground cloves	
2	19-oz. (540-mL) can pinto beans, drained and liquid reserved	2
1	28-oz. (796-mL) can tomatoes, crushed	1
4 cups	winter squash, peeled and cubed	1 L
1 1/2 cups	fresh or frozen corn	360 mL
	salt and freshly ground black pepper to taste	
2 Tbsp.	chopped fresh cilantro (optional)	30 mL

Serves 6

This hearty stew is inspired by the cuisine of Central America. It can be served alone or with steamed rice on the side.

Heat the oil in a large saucepan over medium-high heat. Add the onion and sauté until softened, about 3 minutes. Add the garlic, paprika, chili powder, cumin, coriander, oregano, cinnamon, nutmeg and cloves and sauté for 30 seconds. Add the drained pinto beans, tomatoes and squash. Simmer until the squash is tender, about 30 minutes. Add the reserved bean water if it becomes too dry. Add the corn and continue to cook for 10 minutes longer. Season with salt and pepper.

Ladle into serving bowls and garnish each serving with chopped cilantro.

MOUSSAKA

Serves 9

This classic Greek dish is always worth the effort. For a relaxed dinner, make it a day ahead and reheat it.

2	medium eggplants, unpeeled, cut into 1/2-inch-thick (1.2-cm) slices	2
1/2 cup	olive oil	120 mL
1	large onion, thinly sliced	1
1 cup	finely chopped carrots	240 mL
1/2 cup	finely chopped celery	120 mL
2 cups	sliced mushrooms	480 mL
3	cloves garlic, finely chopped	3
1 tsp.	dried oregano	5 mL
1/2 tsp.	ground cinnamon	2.5 mL
1	28-oz. (796-mL) can of plum tomatoes, crushed	1
1/4 cup	chopped fresh Italian parsley	60 mL
	salt and freshly ground black pepper to taste	
1 cup	grated Parmesan cheese	240 mL
1/3 cup	butter	80 mL
1/4 cup	flour	60 mL
3 cups	cold milk	720 mL
4	egg yolks	4

Sprinkle the eggplant slices with salt, place in a draining rack and let sit for 20 minutes. Rinse with cold water, drain and pat dry with paper towel.

Preheat the oven to 425°F (220°C).

Brush both sides of the eggplant slices with some of the olive oil. Place on baking sheets and bake for 10 minutes. Turn the eggplant and bake for another 15 minutes. Remove from the oven and set aside to cool. Reduce the oven temperature to 350°F (175°C).

Heat the remaining olive oil in a large sauté pan over medium heat. Add the onion, carrots and celery, and sauté until the onions are translucent, approximately 5 minutes. Add the mushrooms and sauté until lightly browned. Add the garlic, oregano and cinnamon, and sauté for 30 seconds. Stir in the tomatoes and parsley. Continue to cook until the sauce begins to thicken, approximately 15 minutes. Season with salt and pepper.

Lightly oil a 9- x 13- x 2-inch (23- x 33- x 5-cm) baking dish. Cover the bottom of the dish with half the eggplant slices. Spoon half the tomato sauce over the eggplant and sprinkle with 1/4 cup (60 mL) of the Parmesan cheese. Repeat the layers with the remainder of the eggplant and sauce and another 1/4 cup (60 mL) of cheese. Set aside.

Melt the butter in a medium saucepan over medium heat. When the butter begins to bubble, whisk in the flour until well combined. Gradually add the milk, stirring continuously with a wooden spoon until the sauce thickens. Stir in 1/4 cup (60 mL) of the cheese, and season with salt and pepper.

Whisk the egg yolks in a large bowl. Very slowly add the hot sauce to the egg yolks, whisking constantly. Pour the sauce over the vegetables in the baking dish. Sprinkle the remainder of the cheese on top.

Bake the moussaka for 45 minutes, or until the sauce is golden brown on top. Let rest 15 minutes before slicing.

MEDITERRANEAN STUFFED SWEET PEPPERS

Serves 4

This recipe calls for sweet red peppers, but for a really colorful entrée, mix red and yellow peppers.

8	sweet red peppers, roasted (see page 5)	8
3 Tbsp.	olive oil	45 mL
1/4 cup	finely chopped shallots	60 mL
2	garlic cloves, finely chopped	2
1 tsp.	crushed red pepper flakes	5 mL
1 tsp.	saffron threads, crushed	5 mL
1 cup	long-grain rice	240 mL
1 cup	chicken stock	240 mL
1/2 tsp.	chopped fresh rosemary	2.5 mL
1/3 cup	dry white wine	80 mL
1 tsp.	olive oil	5 mL
1 cup	marinated artichoke hearts, drained	240 mL
1 cup	stuffed green olives, sliced	240 mL
	salt and freshly ground black pepper to taste	
1 Tbsp.	olive oil	15 mL
1/2 tsp.	chopped fresh rosemary	2.5 mL
1/3 cup	dry white wine	80 mL
8	sprigs fresh rosemary (optional)	8

Set aside the 8 best pepper halves for stuffing. Purée 4 of the remaining pepper halves in a blender and set them aside. Finely dice the remaining 4 pepper halves and set them aside.

Heat the 3 Tbsp. (45 mL) of olive oil over medium heat, add the shallots and sauté lightly for 1 minute. Stir in the reserved diced peppers, garlic, red pepper flakes and saffron. Add the rice, chicken stock, 1/2 tsp. (2.5 mL) rosemary and 1/3 cup (80 mL) white wine. Bring to a boil, cover and reduce the heat. Simmer for 20 minutes, or until all the liquid is absorbed. Remove from the heat. Let the rice rest for 5 minutes, add the 1 tsp. (5 mL) of olive oil and fluff the rice with a fork.

Transfer the rice to a large bowl and toss it with the artichoke hearts and olives. Season with salt and pepper.

Preheat the oven to 350°F (175°C).

Place the reserved peppers in a shallow baking dish. Fill the pepper halves with the rice filling, dividing it equally between them.

Add the 1 Tbsp. (15 mL) of olive oil, 1/2 tsp. (2.5 mL) of rosemary and the remaining 1/3 cup (80 mL) of wine to the pepper purée. Season with salt and pepper. Pour the sauce around the peppers in the baking dish. Cover with aluminum foil and bake for 20 minutes. Remove the foil and bake for another 5 minutes.

To serve, spoon a little sauce on each plate, place a stuffed pepper on top and garnish with a sprig of fresh rosemary.

GRILLED VEGETABLE PAELLA

Serves 4

A hearty Mediterranean dish that makes a very colorful presentation.

2 Tbsp.	olive oil	30 mL
1	clove garlic, finely chopped	1
4	zucchini, 2 yellow and 2 green, cut diagonally into 1/4-inch (.6-cm) slices	4
2	Japanese eggplants, cut diagonally into 1/4-inch (.6-cm) thick slices	2
2	sweet red peppers, quartered	2
1	red onion, cut into 8 wedges	1
2 Tbsp.	olive oil	30 mL
3	cloves garlic, finely chopped	3
1/2 tsp.	curry powder	2.5 mL
1 Tbsp.	fresh rosemary	15 mL
1 tsp.	grated orange zest	5 mL
1 tsp.	grated lemon zest	5 mL
2 cups	Arborio rice	480 mL
6 cups	chicken or vegetable stock	1.5 L
1 tsp.	saffron threads	5 mL
1/2 tsp.	crushed red pepper flakes	2.5 mL
1	8-oz. (227-mL) jar marinated artichoke hearts, drained	1
1/2 cup	stuffed green olives, drained	120 mL
4	green onions, finely chopped	4
1/2 cup	frozen peas, thawed	120 mL
	salt and freshly ground black pepper to taste	
1	fresh sprig rosemary (optional)	1

Combine 2 Tbsp. (30 mL) of olive oil with 1 chopped clove of garlic. Toss lightly with the zucchini, eggplant, red pepper and red onion. Let stand for 2 hours at room temperature.

Grill the marinated vegetables over medium heat until they are just beginning to get tender. (Alternately, place them on a baking sheet in a 350°F (175°C) oven and bake until slightly tender, about 45 minutes). Remove from the heat and set aside.

Heat 2 Tbsp. (30 mL) of olive oil in a large sauté pan over medium heat. Add the 3 cloves of chopped garlic, curry powder and rosemary, and sauté for 30 seconds. Add the orange and lemon zest, rice, stock, saffron threads and red pepper flakes. Bring the mixture to a boil. Reduce the heat and simmer for 10 minutes.

Add the grilled vegetables, artichoke hearts and olives and simmer for 10 minutes, or until the rice is tender. Stir frequently, adding more stock if necessary. Add the green onions and peas, and cook for 3 minutes more. Season with salt and pepper and garnish with a fresh rosemary sprig if desired.

ROASTED BARLEY AND MUSHROOM PILAF

Serves 4

*A nice chewy texture
and nutty flavor best
describes this barley
pilaf. The recipe calls for
white button mushrooms
but any other variety,
such as shiitake or
portobello, makes an
interesting change.*

2 Tbsp.	vegetable oil	30 mL
2 cups	pearl barley	480 mL
2	shallots, finely chopped	2
1/2 lb.	white button mushrooms	225 g
1/2 cup	dry white wine	120 mL
3 1/2 cups	chicken stock	840 mL
2 Tbsp.	butter	30 mL
1/4 cup	grated Parmesan cheese	60 mL
2 Tbsp.	chopped fresh sage or thyme	30 mL
	salt and freshly ground black pepper to taste	
	few fresh sage leaves or thyme sprig (optional)	

Preheat the oven to 350°F (175°C).

Heat the oil in an ovenproof saucepan over medium heat. Add the barley and sauté, stirring often, until it starts to brown, 5 to 8 minutes. Add the shallots and cook for 2 minutes, or until they begin to soften. Do not allow the shallots to brown. Add the mushrooms and continue to cook until wilted, approximately 5 minutes. Stir in the wine and cook until it's completely absorbed, stirring often. Add 1/2 cup (120 mL) of the stock and stir until it is all absorbed.

Add the remaining stock, bring to a boil and cover. Place in the oven and bake until all the stock has been absorbed and the barley is tender, approximately 45 minutes.

Remove from the oven. Stir in the butter and Parmesan cheese and fluff up the pilaf with a fork. Add the chopped sage or thyme and season with salt and pepper. Garnish with fresh sage or a thyme sprig if desired.

WILD MUSHROOM RISOTTO

2 Tbsp.	oil	30 mL
2 Tbsp.	butter	30 mL
1	small onion, finely chopped	1
1 cup	Arborio rice	240 mL
1 cup	dry white wine	240 mL
4 cups	chicken stock, hot	1 L
2 Tbsp.	butter	30 mL
2 cups	sliced white mushrooms	480 mL
1/2 cup	dried wild mushrooms, reconstituted (see page 166)	120 mL
1/2 cup	grated Parmesan cheese	120 mL
2 tsp.	white truffle oil (optional)	10 mL

Serves 2

The secret to successful risotto is continuous stirring. The more you stir, the more starch is released from the rice, making the dish creamier. A word of caution: have extra stock on hand, as the rice may absorb more or less than what's called for in the recipe.

Heat the oil and 2 Tbsp. (30 mL) of butter in a medium sauté pan over medium heat. When the butter is bubbly, add the onion and sauté for 2 minutes. Add the rice, and sauté over low heat for 1 to 2 minutes, until all the grains are coated with oil. Add the wine and continue cooking until it has evaporated. Add the stock a little at a time. Continue to cook, stirring frequently, until most of the stock is absorbed and the rice is tender, approximately 20 minutes.

Meanwhile melt the remaining 2 Tbsp. (30 mL) butter in a separate pan over medium heat. Add the white mushrooms and sauté until lightly browned. Add the wild mushrooms and continue to sauté for 3 minutes, or until heated through.

Add the Parmesan cheese and the mushrooms to the finished risotto. Serve on individual plates. Spoon 1 teaspoon (5 mL) of truffle oil over each serving, if desired.

Spicy Cabbage Rolls

Makes 24 rolls

A traditional Ukrainian favorite, cabbage rolls have been around for generations. They are great supper fare when the weather turns cool, and they're perfect for a buffet table.

1/4 cup	butter	60 mL
1	large onion, finely chopped	1
2/3 cup	barley	160 mL
2/3 cup	rice	160 mL
1	large clove garlic, finely chopped	1
1 tsp.	salt	5 mL
2 1/2 cups	chicken stock	600 mL
1	large head cabbage	1
1/2 cup	pine nuts, toasted (see page 183)	120 mL
1 cup	raisins	240 mL
3 Tbsp.	chopped fresh chives	45 mL
1 1/2 Tbsp.	Dijon mustard	22.5 mL
1/2 tsp.	cumin	2.5 mL
1 tsp.	crushed red pepper flakes	5 mL
	freshly ground black pepper to taste	
1	28-oz. (796-mL) can spicy tomato sauce	1
1 Tbsp.	brown sugar	15 mL
3 cups	sauerkraut	720 mL
2 tsp.	caraway seeds	10 mL
2 cups	sour cream	480 mL

Melt the butter in a saucepan over medium-high heat. Add the onion and sauté until translucent. Add the barley and rice and continue to cook, stirring constantly, until all the grains are coated with butter. Add the garlic and sauté for 30 seconds. Add the salt and stock and bring to a boil. Reduce the heat to low, cover the saucepan with a tight-fitting lid and continue to simmer for 20 minutes, or until the rice and barley are tender. Remove from the heat and let rest for 5 minutes. Fluff with a fork. Set aside to cool.

Bring a large pot of salted water to a boil. Remove the core from the cabbage and discard. Place the cabbage in the water and simmer for 20 minutes. Peel the leaves away one by one from the head as they become tender and set them aside to drain. Cut away the heavy rib from the center of each leaf.

In a large bowl, combine the rice and barley mixture with the pine nuts, raisins, chives, mustard, cumin and red pepper flakes. Mix well and adjust the seasoning with pepper.

Combine the tomato sauce and brown sugar in a bowl and mix well. Set aside.

Preheat the oven to 325°F (165°C).

Place 2 Tbsp. (30 mL) of filling on each cabbage leaf and roll up the leaves, tucking in the sides as you roll. Secure with a toothpick if necessary. Continue until all the filling is used.

Butter a 10- x 15-inch (25- x 38-cm) ovenproof baking dish. Place the unused leftover cabbage leaves on the bottom of the baking dish. Spread 1/2 of the sauerkraut over the cabbage leaves, sprinkle 1 tsp. (5 mL) of caraway on top, and add a little of the tomato sauce.

Place the cabbage rolls on top of the sauerkraut and tomato sauce. Place the rolls tight together in the pan. Spread the remainder of the sauerkraut and caraway seeds on top of the cabbage rolls. Pour the remaining tomato sauce over the top of the rolls and cover tightly with aluminum foil. Bake for 1 hour. Remove the foil and continue to bake for 30 minutes more. Check now and then that they aren't drying out. If they do, just add a little tomato juice.

Serve with lots of sour cream.

Note: If the canned tomato sauce is really thick, thin it a little with tomato juice.

TOMATO, BLUE CHEESE AND SWEET ONION RUSTICA

Serves 4 to 6

This free-form tart is perfect for an informal get-together. Serve it with a crunchy green salad and fresh pears. This recipe makes one 14-inch (36-cm) tart or 4 individual 8-inch (20-cm) tarts.

For the crust:

1/2 cup	lukewarm water	120 mL
1/2 tsp.	sugar	2.5 mL
1 tsp.	active dry yeast	5 mL
1/2 tsp.	salt	2.5 mL
1 1/2 cups	all purpose flour	360 mL
1	large egg, at room temperature	1
3 Tbsp.	butter, softened	45 mL

In a small bowl mix the water and sugar. Sprinkle the yeast on top and set aside to proof for 5 to 10 minutes. In a medium bowl mix the salt and flour. Make a well in the center. Beat the yeast until it's no longer frothy. Place the egg, butter and yeast mixture in the well. Stir until the dough comes together in a soft ball. Knead until smooth, 2 to 3 minutes. Place the ball of dough in an oiled bowl, cover with a damp cloth and set in a warm, draft-free place for 45 minutes.

To make the tart:

3 Tbsp.	olive oil	45 mL
1 1/2 lbs.	onions, thinly sliced	680 g
6	fresh thyme sprigs	6
2 tsp.	chopped fresh rosemary	10 mL
1 tsp.	sugar	5 mL
	salt and freshly ground black pepper to taste	
3 oz.	blue cheese	85 g
6	medium plum tomatoes, cut into 1/4-inch-thick (.6-cm) slices	6
1 Tbsp.	olive oil	15 mL
1	large egg, beaten	1

Heat the 3 Tbsp. (45 mL) of olive oil in a medium sauté pan over medium-low heat. Add the onions, thyme, rosemary, sugar, salt and pepper. Sauté for approximately 15 minutes. Cover and continue to cook for 30 minutes longer, stirring often, until the onions are a caramel-brown color. Be careful not to burn them. Remove from the heat and cool.

Preheat the oven to 400°F (200°C).

Punch the dough down and form it into a ball. Using a rolling pin, roll the dough very thin, making a circle approximately 16 inches (40 cm) in diameter. Transfer the dough to a baking sheet.

Spread the onion in the center, leaving a 2-inch (5-cm) border around the edge of the tart. Crumble the blue cheese over the onion mixture. Place the tomato slices on top of the cheese, overlapping them. Sprinkle with salt and pepper and drizzle with the 1 Tbsp. (15 mL) olive oil. Fold the sides in, pinching the dough to keep it in place. Brush the dough edges with the beaten egg. Bake in the center of the oven for 20 to 30 minutes, or until the crust is golden.

ASPARAGUS AND ROASTED RED PEPPER FLAN

Serves 6

We can never decide whether we like asparagus with red peppers or red peppers with asparagus. Either way they are sensational. This flan is perfect for a brunch gathering; serve it with a tossed salad.

For the pastry:

1 cup	all purpose flour	240 mL
1/2 tsp.	salt	2 mL
1/3 cup	shortening or lard, chilled	80 mL
1 Tbsp.	butter, chilled	15 mL
2 Tbsp.	cold water (more or less)	30 mL

Sift the flour and salt together in a large bowl. Cut in the shortening or lard and butter with a pastry blender until the dough is pea-size. Sprinkle with the water and continue blending until all the ingredients hold together. Gather into a ball, flatten into a disc shape, wrap in plastic and let rest in the refrigerator while working on the flan filling.

To make the flan:

1 cup	asparagus, cut into 1-inch (2.5-cm) pieces	240 mL
1 cup	half-and-half cream, heated	240 mL
1	egg white, beaten	1
1 cup	grated Swiss cheese	240 mL
1/2	sweet red pepper, roasted (see page 5) and cut into 1/4-inch (.6-cm) dice	1/2
3	eggs	3
1/4 tsp.	salt	1.2 mL
1/4 tsp.	freshly ground black pepper	1.2 mL
	pinch cayenne pepper	
	pinch freshly ground nutmeg	

Bring a small pot of salted water to a boil. Add the asparagus, cook for 2 minutes, and plunge into cold water to stop the cooking process. When the asparagus is cooled, drain and set aside.

Preheat the oven to 375°F (190°C). Have ready a 9-inch (23-cm) tart pan with a removable bottom.

Scald the cream to hasten the cooking time. Remove from the heat and cool a little.

Roll out the pastry dough to 2 inches (5 cm) larger than the tart pan. Transfer the dough to the pan. Turn the extra dough inward to form a double thickness of crust around the perimeter. Pinch it together; it should rise about 1/4 inch (.6 cm) above the tart pan. Brush the bottom lightly with the egg white.

Sprinkle half the Swiss cheese over the bottom of the flan. Top with the roasted pepper and asparagus pieces, reserving the tips to garnish the top.

Beat the eggs in a bowl until creamy. Add the warm cream, salt, black pepper, cayenne pepper and nutmeg and mix well. Pour into the flan shell.

Sprinkle the remaining Swiss cheese and the reserved asparagus tips on top. Bake for 35 to 40 minutes, or until the top is golden brown.

WILD MUSHROOM TART

For the pastry:

1 1/4 cups	all purpose flour	300 mL
1/2 tsp.	salt	2.5 mL
1/2 cup	unsalted butter, cut into small pieces and chilled	120 mL
2 Tbsp.	cold water	30 mL

Serves 6

If wild mushrooms are not available, just use button mushrooms. The tart will still be delicious.

Sift the flour and salt together in a large bowl. Cut in the butter with a pastry blender until the dough is pea-size. Sprinkle with the water and continue blending until all the ingredients hold together.

Gather into a ball, flatten into a disc shape, wrap in plastic and let rest in the refrigerator for 1 hour.

Have ready a 9-inch (23-cm) tart pan with a removable bottom.

Roll out the pastry dough so it is 2 inches (5 cm) larger than the tart pan. Transfer the dough to the pan. Turn the extra dough inward to form a double thickness of crust around the perimeter. Pinch it together; it should rise about 1/4 inch (.6 cm) above the tart pan. Chill in the refrigerator for 20 minutes.

RECONSTITUTING MUSHROOMS

To reconstitute dried mushrooms, place the mushrooms in warm water for approximately 20 to 30 minutes. Change the water occasionally. Remove the mushrooms with a slotted spoon, discarding the old water and the sandy debris on the bottom of the bowl.

To make the tart:

1/4 cup	unsalted butter	60 mL
2	shallots, finely chopped	2
1 oz.	dried wild mushrooms, reconstituted and roughly chopped (see facing page)	28 g
2 cups	sliced button mushrooms	480 mL
1 tsp.	finely chopped fresh thyme	5 mL
1 tsp.	finely chopped fresh basil	5 mL
2 Tbsp.	brandy or cognac	30 mL
3/4 cup	milk	180 mL
3	large eggs	3
	pinch freshly grated nutmeg	
	salt and freshly ground black pepper to taste	
1	egg white, beaten	1
1/2 cup	grated Gruyère cheese	120 mL

Melt the butter in a large skillet over medium-high heat. Add the shallots and sauté for 1 minute. Add the wild and button mushrooms and season with salt and pepper. Sauté the mushrooms until golden brown. Stir in the thyme and basil, and add the cognac. Remove from the heat.

Scald the milk in a small saucepan and set aside to cool. Whisk the eggs in a large bowl. Slowly add the cooled milk, whisking continuously. Add the nutmeg, salt and pepper.

Preheat the oven to 375°F (190°C). Lightly brush the chilled tart crust with the beaten egg white. Sprinkle half the grated cheese over the bottom of the tart. Add the mushroom mixture. Pour the custard over the mushrooms and top with the remainder of the grated cheese. Bake until the filling is set and the tart is golden brown, 30 to 45 minutes.

Cool on a rack before serving to allow the filling to firm up. Serve warm.

FLAMICHE LEEK AND BRIE TART

Serves 6

This classic tart of France pairs leeks and Brie cheese perfectly.

For the pastry:

1 cup	all purpose flour	240 mL
1/2 tsp.	salt	2.5 mL
1/3 cup	shortening or lard, chilled	80 mL
1 Tbsp.	butter, chilled	15 mL
2 Tbsp.	cold water	30 mL

Sift the flour and salt together in a large bowl. Cut in the shortening or lard and butter with a pastry blender until the dough is pea-size. Sprinkle with the water and continue blending until all the ingredients hold together. Gather into a ball, flatten into a disc shape, wrap in plastic and let rest in the refrigerator while working on the tart filling.

To make the tart:

2 Tbsp.	butter	30 mL
4	medium leeks, white part only, finely chopped	4
3/4 cup	whipping cream	180 mL
6 oz.	Brie cheese, rind removed and cubed	170 g
	salt and freshly ground black pepper to taste	
1	egg white, beaten	1

Melt the butter in a large sauté pan over medium-high heat. Add the leeks and sauté for 1 minute. Reduce the heat to medium, season with a little salt, and continue to cook until the leeks begin to soften, 3 to 4 minutes. Add the cream and cook for 5 minutes longer. Add more cream if it dries out too much. Add the cheese and cook until it has melted. Remove from the heat and season with salt and pepper.

Preheat the oven to 400°F (200°C). Have ready a 9- or 10-inch (23- to 25-cm) tart pan with a removable bottom.

Roll out the pastry dough so it is 2 inches (5 cm) larger than the tart pan. Transfer the dough to the pan. Turn the extra dough inward to form a double thickness of crust around the perimeter. Pinch it together; it should rise about 1/4 inch (.6 cm) above the tart pan.

Brush the bottom and sides lightly with the egg white. Place the leek filling in the tart pan. Bake for 30 minutes, or until the tart is golden brown on top.

ONION, SPINACH AND PANCETTA PIE

For the pastry:

2 cups	all purpose flour	480 mL
1 tsp.	salt	5 mL
2/3 cup	shortening or lard	160 mL
2 Tbsp.	butter, chilled	30 mL
4 Tbsp.	cold water	60 mL

Serves 6

Serve this pie hot for a light supper or take it along on your next picnic. Served at room temperature, it makes a nice change from the regular picnic fare.

Sift the flour and salt together in a large bowl. Cut in the shortening or lard and butter with a pastry blender until the dough is pea-size. Sprinkle with the water and continue blending until all the ingredients hold together. Gather into a ball. Divide the dough in half and flatten into two discs, cover with plastic wrap and let rest in the refrigerator while working on the pie filling.

To make the pie:

1/2 lb.	pancetta, chopped	225 g
3	large Spanish onions, finely chopped	3
1	10-oz. (285-g) package frozen spinach, thawed, squeezed dry and finely chopped	1
1/2 cup	grated mozzarella cheese	120 mL
1/2 cup	crumbled feta cheese	120 mL
4	eggs, lightly beaten	4
2 Tbsp.	chopped fresh oregano	30 mL
1 tsp.	chopped fresh thyme	5 mL
	salt and freshly ground black pepper to taste	
1	egg, lightly beaten	1
2 tsp.	milk	10 mL

Sauté the pancetta in a large sauté pan over medium-high heat until it is browned. Using a slotted spoon, transfer the pancetta to a large bowl. Drain off all but 2 Tbsp. (30 mL) of the fat from the skillet and reduce the heat to medium. Add the onion and sauté until it's translucent, approximately 5 minutes. Combine the onions and pancetta and let cool slightly while rolling out the pastry.

Preheat the oven to 375°F (190°C). Have ready an 8- or 9-inch (20- or 23-cm) tart pan with 2-inch (5-cm) sides and a removable bottom.

Roll out half the pastry dough so it's 3 inches (7.5 cm) larger than the tart pan. Line the pan with the pastry, trimming it so 1 inch (2.5 cm) hangs over the edge of the tart pan.

Add the spinach, mozzarella and feta cheeses, eggs, oregano, thyme, salt and pepper to the cooled pancetta and onion. Mix well. Transfer to the pastry-lined tart pan.

Mix the egg and milk together. Brush a little of the egg mixture around the edge of the tart dough with a pastry brush.

Roll out the second pastry disc and fit it on the top of the pie. Pinch the top and bottom together. Turn the dough under all around the perimeter to give a finished edge. Crimp it to give the tart a decorative finish. With a sharp knife make gashes in the top to vent the pie. Brush the top with the egg mixture. Bake for 45 minutes to 1 hour, or until the pie is golden brown.

Spinach and Tomato Flan

For the pastry:

1 cup	all purpose flour	240 mL
1/2 tsp.	salt	2.5 mL
1/3 cup	shortening or lard	80 mL
1 Tbsp.	butter, chilled	15 mL
2 Tbsp.	cold water	30 mL

Serves 6

Served hot from the oven or at room temperature, this flan makes a light dinner for warm summer nights.

Sift the flour and salt together in a large bowl. Cut in the shortening or lard and butter with a pastry blender until the dough is pea-size. Sprinkle with the water and continue blending until the dough holds together. Gather into a ball, flatten into a disc shape, wrap in plastic and let rest in the refrigerator while preparing the filling.

To make the flan:

2 Tbsp.	butter	30 mL
1	shallot, finely chopped	1
1	10-oz. (285-g) package frozen spinach, thawed, squeezed dry and chopped salt and freshly ground black pepper to taste	1
1 cup	half-and-half cream	240 mL
3	eggs, beaten pinch freshly grated nutmeg	3
1	egg white, beaten	1
1 cup	grated Gruyère cheese	240 mL
2	medium tomatoes, thinly sliced	2

Melt the butter in a medium sauté pan over medium heat. Add the shallot and sauté until translucent, 1 to 2 minutes. Add the spinach and stir until well combined. Season with salt and pepper. Remove from the heat and set aside.

Scald the cream to hasten the cooking time. Set aside to cool a little.

In a medium bowl, beat the eggs. Add the cooled cream slowly, whisking constantly. Season with nutmeg, salt and pepper.

Preheat the oven to 375°F (190°C). Have ready a 9-inch (23-cm) tart pan with a removable bottom.

Roll out the pastry dough so it's 2 inches (5 cm) larger than the tart pan. Turn the extra dough inward to form a double thickness of crust around the perimeter. Pinch it together; it should rise about 1/4 inch (.6 cm) above the tart pan. Brush the bottom lightly with the beaten egg white.

Sprinkle half the grated cheese over the bottom of the flan. Spread the spinach mixture over the cheese, and pour the egg mixture over the spinach. Place the tomato slices on top, and sprinkle with the remainder of the cheese.

Bake for 30 to 45 minutes, or until the top is golden brown and the filling is set. Let the pie sit 10 to 15 minutes before serving.

Mushroom and Leek Pot Pie

Serves 6 to 8

Vegetable comfort food best describes this delicious pie. This dish can be transformed by adding 2 cups (480 mL) of cooked chicken, shrimp or salmon just before baking and reducing the quantity of mushrooms by half.

1/4 lb.	pearl onions	113 g
	pinch salt	
	pinch brown sugar	
3 Tbsp.	butter	45 mL
1 cup	fresh or frozen peas	240 mL
4 Tbsp.	butter	60 mL
2	carrots, thinly sliced	2
2	leeks, white parts only, thinly sliced	2
2 1/2 lbs.	fresh mushrooms, quartered	1.1 kg
2 Tbsp.	medium dry sherry	30 mL
3 Tbsp.	flour	45 mL
1 1/2 cups	chicken stock	360 mL
1/2 cup	whipping cream	120 mL
1 tsp.	chopped fresh parsley	5 mL
1 1/2 Tbsp.	chopped fresh tarragon	22 mL
	salt and freshly ground black pepper to taste	
1	16-oz. (455-g) package frozen puff pastry, thawed	1
1	egg yolk	1
2 tsp.	milk	10 mL

Peel and remove the ends from the pearl onions. (To make this job easier, place the onions in a bowl of warm water for 10 minutes before peeling.) Place the onions in a small saucepan with water to cover and add the salt and sugar. Cover with a lid, and bring to a boil. Reduce the heat and simmer until tender, approximately 6 minutes. Drain and return to the saucepan. Add the 3 Tbsp. (45 mL) of butter and sauté the onions over medium heat, stirring often, until just golden, 3 to 4 minutes. Remove with a spoon and place in a 10-inch (25-cm) round baking dish with a 2- or 3-quart (2- to 3-L) capacity.

If using frozen peas, place them in the casserole with the onions. If using fresh peas, blanch them in boiling water for 2 minutes, drain and add them to the onions.

Melt the 4 Tbsp. (60 mL) of butter in a large skillet over medium-low heat. Add the carrots and leeks and sauté, covered, for 8 minutes. Stir occasionally, making sure the vegetables do not brown. Remove with a slotted spoon and place in the baking dish with the peas and onions. Add the mushrooms to the pan and sauté until wilted, adding more butter if necessary to prevent the mushrooms from sticking. Place them in the baking dish.

There should be about 2 Tbsp. (30 mL) of drippings in the pan. If not, add a little butter. Return the pan to the heat, add the sherry and reduce until half the liquid has evaporated. Add the flour and stir constantly for 1 minute over medium-high heat. Gradually stir in the stock and then the cream, scraping up any brown bits from the bottom of the pan. Cook for about 5 minutes, or until the sauce is smooth and thick. Stir in the parsley and tarragon. Taste, and adjust seasoning with salt and pepper.

Preheat the oven to 425°F (220°C).

Roll out enough of the pastry dough to fit the top of the baking dish. Depending on the shape of the dish, it may take the whole package. Place on top of the dish and make a few slashes for the steam to escape.

Mix the egg yolk and milk together and brush over the dough. If there is leftover pastry, you can make shapes with it to decorate the crust.

Place the pie in the center of the oven and reduce the temperature to 375°F (190°C). Bake for 40 to 50 minutes, or until the crust is golden brown.

RATATOUILLE

Serves 8

Versatile ratatouille can be served hot as a main dish or as a side dish with roasted meat or fish. It can also be served cold, which makes it easy to bring along on picnics.

3 Tbsp.	olive oil	45 mL
2	large onions, chopped	2
1	large eggplant, cut into 1-inch (2.5-cm) cubes	1
2	zucchinis, cut into 1-inch (2.5-cm) pieces	2
1	sweet red pepper, cut into 1-inch (2.5-cm) pieces	1
1	sweet green pepper, cut into 1-inch (2.5-cm) pieces	1
4	cloves garlic, minced	4
1/2 tsp.	crushed hot pepper flakes	2.5 mL
10	large tomatoes, seeded and chopped	10
1 tsp.	salt	5 mL
1/2 tsp.	saffron threads	2.5 mL
2 tsp.	fresh thyme	10 mL
1 tsp.	chopped fresh rosemary	5 mL
1	bay leaf	1
1/4 cup	chopped fresh basil	60 mL
	salt and freshly ground black pepper to taste	

Heat the oil in a heavy saucepan over medium-high heat. Add the onions and sauté until translucent. Add the eggplant and sauté for 5 minutes, or until it begins to brown. Remove from the pan.

Add more oil if necessary, then add the zucchini and peppers and sauté for 5 minutes. Add the garlic and hot pepper and sauté for 30 seconds. Add the tomatoes, salt, saffron, thyme, rosemary, bay leaf and onion-eggplant mixture. Cover and continue to cook over medium heat until the vegetables are tender, approximately 45 minutes.

Just before serving, discard the bay leaf and stir in the fresh basil. Adjust the seasoning with salt and pepper.

Note: Ratatouille can be refrigerated and served the next day.

Zucchini and Lemon over Linguine

1/2 cup	dry white wine	120 mL
1/4 tsp.	saffron threads	1.2 mL
2 Tbsp.	olive oil	30 mL
2	medium zucchini, cut into 1/4- x 2-inch (.6- x 5-cm) julienne strips	2
	salt and freshly ground black pepper to taste	
1 Tbsp.	butter	15 mL
3	large shallots, thinly sliced	3
2	cloves garlic, finely minced	2
1 cup	chicken stock	240 mL
1/2 cup	whipping cream	120 mL
2 tsp.	grated lemon zest	10 mL
12 oz.	dried linguine	340 g
1/4 cup	grated Parmesan cheese	60 mL

Serves 4

Lemons and zucchini are paired in this easy and tasty pasta dish. It looks lovely with thin slices of lemon as a garnish.

Warm the wine in a small saucepan over low heat. Remove from the heat, add the saffron and set aside.

Heat the olive oil in a large sauté pan over medium heat. Add the zucchini and sauté until it starts to brown. Transfer to a plate, season with salt and pepper and keep warm.

In the same skillet, melt the butter. Add the shallots and sauté over medium heat until translucent, about 4 minutes. Add the garlic and sauté for 30 seconds. Raise the heat, add the wine and stock and bring to a boil. Cook until the liquid is reduced by half. Add the cream and reduce by half again. Stir in the lemon zest and season with salt and pepper.

Meanwhile, bring a large pot of salted water to a boil. Add the linguine to the boiling water and cook until just tender but still firm, 8 to 10 minutes. Drain the pasta, return it to the pot, add the sauce and toss. Transfer to a serving bowl, add the Parmesan cheese and half the zucchini and toss gently. Arrange the remainder of the zucchini on top and serve.

GRILLED VEGETABLE LASAGNE

Serves 9

This recipe calls for grilled vegetables, but oven baking would work just as well. If you're in a hurry, substitute commercially prepared tomato sauce for the Arrabbiata.

8 cups	Arrabbiata Sauce (page 180)	2 L
12	dried lasagne noodles	12
1/4 cup	vegetable oil	60 mL
2	cloves garlic, finely chopped	2
1/2	small eggplant, cut into 1/4-inch-thick (.6-cm) slices, approximately 12 slices	1/2
1/2	Spanish onion, cut into 1/4-inch-thick (.6-cm) slices	1/2
4	carrots, peeled and sliced in half lengthwise	4
4	medium sweet red peppers, quartered	4
1	medium zucchini, cut diagonally into 1/4-inch-thick (.6-cm) slices	1
2 Tbsp.	vegetable oil	30 mL
2 cups	sliced mushrooms	480 mL
1	10-oz. (285-g) pkg. frozen spinach, thawed, drained and chopped	1
3 cups	grated Mozzarella cheese	720 mL
1 cup	grated Parmesan cheese	240 mL
2 cups	ricotta or cottage cheese	480 mL

Heat the sauce and keep it warm while preparing the rest of the ingredients.

Prepare the lasagne noodles according to package instructions. Drain and set aside.

Preheat the barbecue grill to medium.

Combine the 1/4 cup (60 mL) vegetable oil and garlic. Baste the eggplant, onion, carrots, red peppers and zucchini. Grill the vegetables until just barely tender. Don't char them. If the red peppers become charred, remove and discard the skins.

Heat the 2 Tbsp. (30 mL) of oil in a medium sauté pan over medium-high heat. Add the mushrooms and sauté until they begin to brown. Set aside.

Preheat the oven to 350°F (175°C). Oil a 9- x 13-inch (23- x 33-cm) baking dish.

Ladle 1 cup (240 mL) of the sauce onto the bottom of the baking dish. Place 4 lasagne noodles over the sauce. Ladle 1 cup (240 mL) of sauce over the noodles. Arrange the eggplant slices, onions and zucchini over the noodles. Sprinkle with 1 cup (240 mL) of mozzarella cheese and 1/3 cup (80 mL) of Parmesan cheese. Top with the ricotta or cottage cheese.

For the second layer, place 4 noodles over the cheeses. Distribute the spinach, red peppers, carrots and mushrooms evenly over the noodles. Ladle 2 cups (480 mL) of sauce on top. Sprinkle with 1 cup (240 mL) of mozzarella cheese and 1/3 cup (80 mL) of Parmesan cheese.

For the third layer, place 4 noodles over the cheeses. Ladle the remainder of the sauce on top, and sprinkle with the remaining mozzarella and Parmesan cheese.

Cover with foil and bake for 45 minutes. Remove the foil, reduce the heat to 300°F (150°C) and bake 30 minutes longer. Let stand for 15 minutes before serving.

Note: To bake the vegetables in the oven, combine all the vegetables in a large roasting pan, coat with the oil and garlic mixture and roast at 325°F (165°C) for 45 minutes to 1 hour, turning frequently.

SPAGHETTI WITH ARRABBIATA SAUCE

Serves 8

This zippy all-vegetable tomato sauce is a quick and basic sauce. Use it any time you have a recipe requiring tomato sauce. Pass around freshly grated Parmesan cheese to serve with the spaghetti.

1	onion, finely chopped	1
1	zucchini, finely chopped	1
1	carrot, finely chopped	1
1	sweet red pepper, finely chopped	1
1/2	small eggplant, finely chopped	1/2
2 Tbsp.	olive oil	30 mL
2	cloves garlic, finely chopped	2
2 Tbsp.	tomato paste	30 mL
1 Tbsp.	chopped fresh basil	15 mL
1 tsp.	chopped fresh thyme	5 mL
1	bay leaf	1
1 tsp.	red pepper flakes	5 mL
2	28-oz. (796-mL) cans plum tomatoes	2
	salt and freshly ground black pepper to taste	
24 oz.	dried spaghetti	680 g

Processing them one at a time, chop the onion, zucchini, carrot, sweet red pepper and eggplant in a food processor. Place each item in a separate bowl after processing. Pulse the tomatoes in the processor to roughly chop them. Do not purée.

Heat the oil in a large saucepan over medium heat. Add the onion and carrot and sauté until the onions begin to turn translucent. Add the zucchini, red pepper and eggplant, and sauté for 3 minutes. Add the garlic and continue to cook for 30 seconds more.

Add the tomato paste, basil, thyme, bay leaf, red pepper flakes and canned tomatoes. Season with salt and pepper. Bring to a boil. Reduce the heat and simmer for at least 15 to 20 minutes, or up to 1 hour.

Meanwhile, bring a large pot of salted water to a boil. Add the spaghetti and cook until just tender but still firm, 8 to 10 minutes. Drain the pasta, divide it evenly among individual serving bowls and ladle the sauce on top.

Note: This recipe makes approximately 8 cups (2 L) of sauce and it freezes well.

FETTUCCINE WITH PIQUANT TOMATO PEPPER SAUCE

1/4 cup	olive oil	60 mL
2	sweet red peppers, chopped	2
3	cloves garlic, chopped	3
5	ripe tomatoes, seeded and chopped	5
1	dried red chili pepper	1
1/4 cup	hazelnuts, toasted and skins removed (see page 183)	60 mL
1 Tbsp.	red wine vinegar	15 mL
1 tsp.	brown sugar	5 mL
1/2 tsp.	salt	2.5 mL
24 oz.	dried fettuccine	680 g

Serves 8

Fettuccine is called for in the recipe, but this lively sauce also works well as a taco sauce, lasagne filling or pizza sauce. Try using it cold as a dip with endive for a vegetarian appetizer.

Heat the oil in a medium saucepan over medium heat. Add the red pepper and sauté for 1 minute. Add the garlic and sauté for 30 seconds. Add the tomatoes and chili pepper and continue to cook for 5 minutes.

Grind the hazelnuts to a fine paste in a food processor. Add the nuts, vinegar, sugar and salt to the sauce. Continue to cook for 5 minutes more. Remove the sauce from the heat and purée in a food processor or blender until smooth.

Meanwhile, bring a large pot of salted water to a boil. Add the fettuccine and cook until just tender but still firm, 8 to 10 minutes. Drain the pasta, divide it evenly among individual serving bowls and ladle the sauce on top.

CHÈVRE AND ROASTED RED PEPPER CREAM SAUCE OVER PENNE

Serves 6

I use this recipe in my cooking classes, and it is a favorite among my students. Once you've tried it you'll see why.

6	sweet red peppers, roasted (see page 5)	6
3 Tbsp.	butter	45 mL
3	cloves garlic, finely chopped	3
1/4 cup	dry sherry	60 mL
2 Tbsp.	chopped fresh basil	30 mL
1 1/2 cups	whipping cream	360 mL
1/2 cup	grated Parmesan cheese	120 mL
	salt and freshly ground black pepper to taste	
6 cups	dried penne	1.5 L
2	sweet red peppers, roasted and cut into 1/2-inch-wide (1.2-cm) strips	2
1/2 cup	chopped fresh basil	120 mL
10 oz.	chèvre, crumbled	285 g
1/4 cup	pine nuts, toasted (see page 183)	60 mL
1/4 cup	grated Parmesan cheese	60 mL
6	fresh basil leaves (optional)	6

Place the 6 roasted red peppers in a food processor and process until puréed. Set aside.

Melt the butter in a medium sauté pan over medium heat. Add the garlic and sauté for 30 seconds. Add the sherry, and continue to cook until the liquid is reduced by half. Add the puréed peppers and the 2 Tbsp. (30 mL) of basil, and continue to sauté for 1 minute more, or until the liquid starts to evaporate. Add the cream and bring to a boil. Remove from the heat, add the 1/2 cup (120 mL) of Parmesan cheese and season with salt and pepper.

Meanwhile, bring a large pot of salted water to a boil. Add the penne and cook until just tender but still firm, 8 to 12 minutes. Drain the pasta, and transfer it to a large bowl. Toss the pasta with the red pepper strips, 1/2 cup (120 mL) basil, chèvre and pine nuts.

Divide the pasta mixture evenly among 6 plates. Ladle a little sauce over each serving, and sprinkle a little Parmesan cheese over top. Garnish with a fresh basil leaf, if desired.

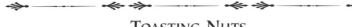

TOASTING NUTS

To toast nuts such as pine nuts, walnuts, almonds, pecans and pistachios, place them in a heavy frying pan over medium-high heat. Stir the nuts frequently until they're lightly toasted on all sides, about 2 to 4 minutes. Watch that they don't burn.

To toast hazelnuts, place them on a baking sheet in a 350°F (175°C) oven for 5 to 8 minutes, or until the skins begin to loosen and the nuts are lightly browned. Stir the nuts occasionally to be sure they are not browning too quickly. To remove the skins, transfer the nuts to a tea towel and rub them briskly.

WOOD-GRILLED PIZZAS WITH PESTO

Serves 4

This authentic wood-grilled pizza recipe has all the classic Italian favorites—basil, red peppers and artichoke hearts—and it can all be made from beginning to end in less than an hour.

For the pizza dough:

2 1/2 cups	all purpose flour	600 mL
1 Tbsp.	quick-rising active dry yeast (1 envelope)	15 mL
1 Tbsp.	sugar	15 mL
1 tsp.	salt	5 mL
2 Tbsp.	olive oil	30 mL
1 cup	lukewarm water	240 mL

Set aside 1 cup (240 mL) of the flour. Combine the remaining flour, yeast, sugar and salt in a large mixing bowl. Make a well in the center. Pour the oil and water into the well. Mix, adding enough of the reserved flour to make a soft dough that does not stick to the bowl.

Turn onto a lightly floured work surface and knead for 8 to 10 minutes, or until the dough is smooth and elastic. Form into a ball and place in a well-oiled bowl. Cover with plastic wrap and let rest for 10 minutes while you prepare the toppings.

To make the pizza:

1 Tbsp.	olive oil	15 mL
1/2 cup	Basil Pesto (page 216)	120 mL
2	sweet red peppers, roasted (see page 5) and sliced	2
1	4-oz. (120-mL) jar marinated artichoke hearts, drained	1
1/2 cup	grated Parmesan cheese	120 mL
1/2 cup	grated mozzarella cheese	120 mL

Preheat the barbecue to medium.

Place approximately 2 cups (480 mL) of hardwood chips in aluminum foil. Spray lightly with water and place on the right side of the barbecue. The barbecue is ready when the chips are smoking lightly.

Punch down the dough and roll it out to fit a 12-inch (30-cm) pizza pan. Lightly oil the pan and place the pizza dough in the pan. Make a border around the edge by pinching the dough. Brush the top with the olive oil. Distribute the pesto evenly over the top. Arrange the sliced peppers and artichoke hearts on top and sprinkle with both cheeses.

Turn off the burner on the left side of the barbecue and place the pizza on that side of the grill. Close the cover and cook for 10 minutes. After 10 minutes, quickly turn the pan and continue to cook for 5 to 10 minutes longer. If the pizza is too light on the bottom, slide it from the pan onto the grill for a minute or two longer. Serve immediately.

QUESADILLAS

For the tortillas:

3 1/2 cups	all purpose flour	840 mL
1 1/2 tsp.	baking powder	7.5 mL
1 tsp.	salt	5 mL
1/3 cup	shortening	80 mL
1 1/4 cups	water	300 mL

Makes 8

You'll be surprised how easy it is to make your own tortillas. Serve these quesadillas with Southwest Tomato Salsa (page 217) and sour cream or our Guacamole (page 8).

Combine the flour, baking powder and salt in a large bowl. Cut in the shortening with a pastry blender until the dough is pea-size. Sprinkle with the water and continue blending until a soft dough is formed.

Turn the dough onto a lightly floured work surface and knead just until the dough comes together in a tidy ball. Divide the dough into 8 pieces and roll each piece into a ball. Cover with a damp cloth and let rest for 30 minutes.

Roll out each ball of dough with a rolling pin to form a 10-inch (25-cm) circle. Heat a lightly oiled frying pan to medium-high. Cook the tortillas one at a time for 1 to 2 minutes per side, or until lightly browned. Stack the tortillas as they are cooked and wrap them in foil to keep them soft.

To make the quesadillas:

4 oz.	cream cheese	113 g
4 oz.	chèvre	113 g
1/4 cup	sun-dried tomatoes, reconstituted (see page 7) and chopped	60 mL
1/4 cup	fresh basil leaves, packed, finely chopped	60 mL
2	green onions, finely sliced	2
2	fresh tomatoes, thinly sliced	2
1 cup	shredded Monterey Jack or Cheddar cheese	240 mL
1	egg white, beaten	1

Combine the cream cheese and chèvre in a food processor. Stir in the sun-dried tomatoes and basil.

Brush one side of a tortilla with oil. Place the tortilla oil side down on a baking sheet. Spread half the tortilla with 1/8 of the cheese mixture, leaving a border around the edge. Top with a few green onions, tomato slices and 1/8 of the shredded cheese. Brush egg white along the outside edge of the tortilla, fold it in half and pinch the edges together. Repeat with all the tortillas.

Cook the quesadillas in a lightly oiled frying pan over medium heat. Cook both sides until lightly browned, approximately 3 minutes per side.

BREADS, LOAVES, MUFFINS AND DESSERTS

Escalated to new heights, vegetables have found their way into breads, muffins, cakes and pies.

The following recipes—from chocolate zucchini muffins to sweet potato pie—may inspire a few more vegetable converts.

MEDITERRANEAN SPINACH AND FETA STUFFED BREAD

For the dough:

1 Tbsp.	active dry yeast (1 envelope)	15 mL
1 cup	lukewarm water	240 mL
3 cups	all purpose flour	720 mL
1 tsp.	salt	5 mL
4 Tbsp.	olive oil	60 mL

Serves 8

Stuffed breads are a lot of fun to make and they look so attractive. This loaf looks sensational when sliced and served with a side salad. The center is green and red, with hints of pine nuts and feta cheese, all surrounded by a delicious bread.

Sprinkle the yeast over the warm water and set aside for 10 minutes to proof.

Combine the flour and salt in a large mixing bowl. Make a well in the center. Stir the frothy yeast mixture and pour it into the well. Add the 4 Tbsp. (60 mL) olive oil. Stir until it's well incorporated, then turn it onto a floured work surface and knead for 8 to 10 minutes, or until it's smooth and elastic.

Form into a ball and place in a well-oiled bowl. Cover with plastic wrap and set aside in a warm, draft-free place for 1 hour, or until doubled in volume.

To make the bread:

8 cups	fresh spinach loosely packed, washed and stems removed, or 1 10-oz. (285-g) pkg. frozen chopped spinach, thawed and squeezed to remove moisture	2 L
1/4 cup	sun-dried tomatoes, reconstituted (see page 7) and chopped	60 mL
1/4 cup	pine nuts, toasted (see page 183)	60 mL
1/4 cup	feta cheese, crumbled	60 mL
	freshly ground black pepper to taste	
1 Tbsp.	olive oil	15 mL

In a large saucepan bring 1 inch (2.5 cm) of salted water to a boil. Add the spinach and steam for approximately 3 minutes, or until the spinach has wilted. Drain well in a colander and squeeze in a towel to remove excess moisture. Chop the spinach. Combine the spinach, sun-dried tomatoes, pine nuts, feta cheese and black pepper in a large bowl. Mix well and set aside.

Roll the dough out on a floured work surface into a 15- x 12-inch (38- x 30 cm) rectangle. Transfer the dough to a well-oiled baking sheet. Distribute the spinach mixture evenly over the dough, leaving a 1-inch (2.5-cm) border around the edges. Sprinkle generously with black pepper and fold the dough in half lengthwise. Pinch the edges to seal. Cover the loaf with a tea towel and set aside to rise for 30 minutes.

Preheat the oven to 400°F (200°C).

Brush the dough with the remaining olive oil and bake for 20 minutes, or until golden brown. Serve warm or at room temperature.

PICANTE BUTTERNUT SQUASH CORNBREAD

Makes 12 muffins

Make this cornbread in muffin tins, or if you're lucky enough to have an old-fashioned cast-iron corn-stick pan, then this is the perfect recipe for it.

1 cup	yellow cornmeal	240 mL
1 cup	all purpose flour	240 mL
1/4 cup	brown sugar	60 mL
2 tsp.	baking powder	10 mL
1 tsp.	salt	5 mL
1/2 tsp.	baking soda	2.5 mL
1/2 cup	grated Cheddar cheese	120 mL
1	egg	1
3/4 cup	buttermilk	180 mL
2 Tbsp.	vegetable oil	30 mL
3/4 cup	butternut squash, cooked and mashed	180 mL
1/3 cup	finely diced sweet red pepper	80 mL
1/2 tsp.	red pepper flakes	2.5 mL

Preheat the oven to 425°F (220°C). Lightly oil a 12-cup muffin pan.

Combine the cornmeal, flour, brown sugar, baking powder, salt, baking powder and cheese in a large bowl.

Beat the egg with the buttermilk, oil and squash. Add the egg mixture to the cornmeal mixture. Add the diced red pepper and red pepper flakes, stirring only until combined. Do not overmix.

Divide the batter evenly among the muffin cups. Bake 15 to 20 minutes, or until golden brown. Serve warm.

Zucchini Lemon Loaf

2 cups	all purpose flour	480 mL
1 cup	whole-wheat flour	240 mL
1 tsp.	baking soda	5 mL
1 tsp.	salt	5 mL
1/2 tsp.	baking powder	2.5 mL
1 1/2 tsp.	ground cinnamon	7.5 mL
1/4 tsp.	allspice	1.2 mL
1/4 tsp.	ground nutmeg	1.2 mL
1 cup	granulated sugar	240 mL
1 cup	chopped walnuts	240 mL
1/2 cup	raisins	120 mL
3	eggs	3
1/2 cup	packed brown sugar	120 mL
1 cup	vegetable oil	240 mL
1 Tbsp.	lemon zest	15 mL
1 tsp.	vanilla	5 mL
2 cups	grated zucchini	480 mL

Makes 2 loaves

A healthy and delicious way to use up all that extra zucchini from the garden.

Preheat the oven to 350°F (175°C). Lightly oil two 4- x 8-inch (10- x 20-cm) loaf pans.

Sift the flours, baking soda, salt, baking powder, cinnamon, allspice, nutmeg and granulated sugar in a large bowl. Stir in the walnuts and raisins.

In a separate bowl, beat the eggs lightly. Add the brown sugar, oil, lemon zest and vanilla and stir until well combined. Add the grated zucchini and mix thoroughly.

Add the wet ingredients to the dry ingredients, stirring just until combined. Divide the batter evenly between the prepared loaf pans. Bake until well browned, approximately 40 to 45 minutes, or until a tester inserted in the center comes out clean.

Stored in plastic wrap, this loaf will keep for up 2 days. Quick breads like this one also freeze well.

Pumpkin Spice Bread with Orange Glaze

Makes 1 loaf

This bread calls for pumpkin, but you can substitute cooked butternut or any other winter-type squash.

2 cups	flour	480 g
2 tsp.	baking powder	10 mL
1 tsp.	ground cinnamon	5 mL
1/2 tsp.	salt	2.5 mL
1/2 tsp.	ground nutmeg	2.5 mL
1/4 tsp.	ground ginger	1.25 mL
1/4 tsp.	ground cloves	1.25 mL
1/4 tsp.	baking soda	1.25 mL
1/4 cup	granulated sugar	60 mL
1/2 cup	shelled pistachios (or walnuts)	120 mL
1/4 cup	raisins	60 mL
2	eggs	2
1/2 cup	packed brown sugar	120 mL
1 cup	cooked pumpkin	240 mL
1/2 tsp.	vanilla extract	2.5 mL
1/3 cup	butter, melted	80 mL
1/3 cup	granulated sugar	80 mL
1 tsp.	grated orange rind	5 mL
1/3 cup	orange juice	80 mL

Preheat the oven to 350°F (175°C). Lightly oil a 5- x 9-inch (12.5- x 23-cm) loaf pan.

Sift the flour, baking powder, cinnamon, salt, nutmeg. ginger, cloves, baking soda and 1/4 cup (60 mL) granulated sugar in a large bowl. Mix in the pistachios and raisins.

Beat the eggs. Add the brown sugar and continue to beat until well combined. Add the pumpkin, vanilla and butter and mix well. Add the pumpkin mixture to the flour mixture, stirring until just combined.

Place the batter in the prepared loaf pan. Bake until the loaf is well browned and a tester comes out clean, approximately 50 minutes.

Meanwhile, heat the 1/3 cup (80 mL) sugar, orange rind and orange juice over medium heat until the sugar is dissolved. Remove from the heat and cool slightly.

Remove the cake from the oven. Cool for 10 minutes. Using a skewer, pierce the loaf all the way to the bottom 12 to 16 times. Loosen the loaf from the sides of the pan. Pour the glaze over it. Let cool completely before removing from the pan.

Stored in plastic wrap, this loaf will keep for up to 2 days. Quick breads like this one also freeze well.

SWEET POTATO CHEDDAR AND CHIVE BISCUITS

Makes approximately 12 biscuits

These biscuits are perfect with a hot bowl of homemade soup. They are best served the same day, but the dough can be made ahead and stored in the fridge for up to 1 day. Cover the rolled and cut biscuit dough with a damp cloth and pop them in the oven when you're ready to bake them.

2 cups	*all purpose flour*	*480 mL*
4 tsp.	*baking powder*	*20 mL*
1 tsp.	*salt*	*5 mL*
2/3 cup	*vegetable shortening*	*160 mL*
1 cup	*cooked sweet potato, mashed*	*240 mL*
3 Tbsp.	*milk (or more)*	*45 mL*
1/2 cup	*grated Cheddar cheese*	*120 mL*
1/4 cup	*finely chopped fresh chives*	*60 mL*

Preheat the oven to 400°F (200°C).

Sift the flour, baking powder and salt together in a large bowl. Cut in the shortening and sweet potato with a pastry blender. Add enough of the milk to make a soft dough. Add half of the Cheddar cheese and all the chives. Gather the dough into a ball and knead for 1 minute.

Roll the dough on a lightly floured surface into a 1-inch-thick (2.5-cm) rectangle. Cut the dough into 2 1/2-inch (6-cm) squares, or cut into rounds with a biscuit cutter. Place the biscuits on an ungreased baking sheet and sprinkle the tops with the remaining Cheddar cheese. Bake for 15 to 20 minutes, or until the cheese is melted and the biscuits are golden brown.

CARROT AND SUNFLOWER SEED MUFFINS

2 cups	all purpose flour	480 mL
1 cup	sugar	240 mL
2 tsp.	baking soda	10 mL
2 tsp.	ground cinnamon	10 mL
1/2 tsp.	salt	2.5 mL
2 cups	grated carrots	480 mL
1/2 cup	raisins	120 mL
1/2 cup	raw, unsalted sunflower seeds	120 mL
1/2 cup	shredded unsweetened coconut	120 mL
1	apple, peeled, cored and grated	1
3	eggs, beaten	3
1 cup	vegetable oil	240 mL
2 tsp.	vanilla	10 mL

*Makes 12
large muffins*

*These could be called
everything-but-the-
kitchen-sink muffins.
The recipe calls for
all purpose flour but
substituting whole wheat
flour for half the all
purpose flour makes
them even more healthy.*

Preheat the oven to 350°F (175°C). Lightly oil 12 large-size muffin cups.

Sift the flour, sugar, baking soda, cinnamon and salt together in a large bowl.

In a medium bowl mix the grated carrot, raisins, sunflower seeds, coconut and apple. Add the fruit mixture to the dry ingredients and mix until well combined.

Mix the eggs, vegetable oil and vanilla together and add to the muffin mixture. Stir only until combined; do not overmix. Spoon the batter into the prepared muffin cups and bake for 20 minutes, or until a tester inserted into the center of the muffins comes out clean. Cool on wire racks.

Muffins can be stored in airtight containers for 2 days; they also freeze well.

CHOCOLATE ZUCCHINI MUFFINS

Makes 12 large muffins

Nobody will believe these rich and chocolatey muffins have a healthy zucchini base.

2 cups	all purpose flour	480 mL
1/4 cup	cocoa powder	60 mL
1 tsp.	baking soda	5 mL
1 1/2 tsp.	baking powder	7.5 mL
1 tsp.	ground cinnamon	5 mL
1 tsp.	ground cloves	5 mL
1/2 tsp.	salt	2.5 mL
1/2 cup	chocolate chips	120 mL
3/4 cup	vegetable oil	180 mL
1 cup	sugar	240 mL
2	eggs, beaten lightly	2
1/2 cup	buttermilk	120 mL
2 cups	grated zucchini	480 mL

Preheat the oven to 350°F (175°C). Lightly oil 12 large-size muffin cups.

Sift the flour, cocoa powder, baking soda, baking powder, cinnamon, cloves and salt together in a large bowl. Stir in the chocolate chips.

In a separate bowl combine the vegetable oil, sugar, eggs, buttermilk and zucchini.

Add the wet ingredients to the dry ingredients, stirring just until combined. Spoon the batter into the prepared muffin pans and bake for 18 to 20 minutes, or until a tester inserted into the center of the muffins comes out clean. Cool on wire racks.

Muffins can be stored in airtight containers for 2 days; they also freeze well.

BUTTERNUT SQUASH AND CRANBERRY MUFFINS

2 1/4 cups	all purpose flour	535 mL
1 cup	sugar	240 mL
1 tsp.	ground cinnamon	5 mL
1/4 tsp.	mace	1.2 mL
1/8 tsp.	ground cloves	.5 mL
1/4 tsp.	nutmeg	1.2 mL
2 tsp.	baking powder	10 mL
1 tsp.	baking soda	5 mL
1/2 tsp.	salt	2.5 mL
1 cup	cooked and mashed or puréed butternut squash	240 mL
1	egg, beaten	1
3/4 cup	vegetable oil	180 mL
1/2 cup	milk	120 mL
1 cup	coarsely chopped cranberries	240 mL

Makes 12 large muffins

Just when you think the summer's harvest is over, there is still some of nature's bounty to be enjoyed. This recipe calls for butternut squash, but Hubbard or acorn squash are just as flavorful. Fresh or frozen cranberries give equally great results.

Preheat the oven to 375°F (190°C). Lightly oil 12 large muffin cups.

Combine the flour, sugar, spices, baking powder, baking soda and salt in a large mixing bowl.

In a separate bowl, combine the mashed squash, egg, oil and milk. Mix well.

Add the wet ingredients to the dry ingredients and mix just until combined; do not overmix. Fold in the cranberries. Spoon the batter into the prepared muffin cups. Bake for 16 to 20 minutes, or until a tester inserted into the center of the muffins comes out clean. Cool on wire racks.

Muffins can be stored in airtight containers for 2 days; they also freeze well.

BUTTERNUT SQUASH CAKE WITH CREAM CHEESE FROSTING

Serves 12 to 16

This cake is like carrot cake or pumpkin loaf, but the squash gives it a lighter color and slightly different flavor.

2 cups	butternut squash, peeled and finely grated	480 mL
3/4 cup	crushed pineapple, well drained	180 mL
1 cup	raisins, washed and dried	240 mL
1 Tbsp.	grated orange zest	15 mL
3/4 cup	finely shredded coconut	180 mL
1/2 cup	pecans, toasted (see page 183) and chopped	120 mL
4	eggs	4
1 cup	honey	240 mL
1/2 cup	brown sugar, packed	120 mL
3/4 cup	vegetable oil	180 mL
2 cups	all purpose flour	480 mL
2 tsp.	baking powder	10 mL
1 tsp.	baking soda	5 mL
1/2 tsp.	salt	2.5 mL
1 Tbsp.	ground cinnamon	15 mL
	pinch allspice	
1/4 tsp.	ground nutmeg	1.2 mL
4 oz.	cream cheese	113 g
2 Tbsp.	butter	30 mL
1 tsp.	vanilla	5 mL
2 tsp.	grated orange zest	10 mL
1 1/2 cups	icing sugar	360 mL

Preheat the oven to 325°F (165°C). Lightly oil a 10-inch (25-cm) bundt pan.

Place the squash, pineapple, raisins, orange zest, coconut and pecans in a large bowl.

Using an electric mixer, beat the eggs, honey and brown sugar until light. Add the oil and continue to beat until well mixed.

In a separate large bowl, sift the flour, baking powder, baking soda, salt and spices. Stir well.

Make a well in the flour mixture. Add the egg mixture and stir until just combined. Add the grated squash mixture. Stir just until all the ingredients are incorporated into the batter. Do not overmix.

Spread the batter evenly into the prepared bundt pan. Bake for 60 to 75 minutes, or until a tester comes out clean. Cool on a rack in the pan.

With an electric mixer beat the cream cheese and butter until smooth. Add the vanilla and orange zest. Gradually add the icing sugar until the icing is smooth and spreadable.

Ice the cake when it is completely cooled.

CLASSIC CARROT CAKE

Serves 10 to 12

Who doesn't love carrot cake? This spicy moist cake has become a standard in the cook's repertoire.

4	eggs	4
1 cup	brown sugar	240 mL
1/2 cup	granulated sugar	120 mL
1/4 cup	honey	60 mL
1 cup	vegetable oil	240 mL
2 cups	all purpose flour	480 mL
2 tsp.	baking powder	10 mL
1 tsp.	salt	5 mL
1/2 tsp.	baking soda	2.5 mL
1 1/2 tsp.	ground cinnamon	7.5 mL
1/2 tsp.	ground nutmeg	2.5 mL
2 cups	grated raw carrots	480 mL
1 cup	crushed pineapple, drained	240 mL
1/2 cup	flaked coconut	120 mL
1/2 cup	raisins	120 mL
1/2 cup	raw sunflower seeds	120 mL
1 tsp.	grated orange zest	5 mL
1 tsp.	grated lemon zest	5 mL
4 oz.	cream cheese	113 g
2 Tbsp.	butter	30 mL
1 tsp.	vanilla	5 mL
1 1/2 cups	icing sugar	360 mL

Preheat the oven to 350°F (175°C). Butter a 10-inch (25-cm) bundt pan and dust with flour.

Beat the eggs with an electric mixer. Add the sugars and honey and continue to beat until light and creamy. Gradually mix in the oil.

Sift together the flour, baking powder, salt, baking soda, cinnamon and nutmeg. Combine with the egg mixture. Add the carrots, pineapple, coconut, raisins, sunflower seeds and orange and lemon zests. Mix well. Pour the batter into the prepared bundt pan. Bake for 60 to 75 minutes, or until a tester comes out clean.

Remove from the oven. Let rest 10 minutes before removing from the pan. Cool on a rack.

Using an electric mixer, beat the cream cheese, butter and vanilla together. Gradually add the icing sugar, beating until smooth and spreadable, but still a little soft. Adjust the quantity of icing sugar to achieve the desired consistency. Frost the top of the cooled cake, leaving the sides unfrosted.

Note: To flavor the icing, add 1 tsp. (5 mL) of lemon or orange zest.

PUMPKIN CHEESECAKE WITH SOUR CREAM TOPPING

Serves 12 to 16

This rich cheesecake is made even spicier with the addition of a crushed gingersnap crust. To get nice smooth pieces, dip the knife in hot water and wipe the blade with a towel, then slice the cake.

For the crust:

1 1/2 cups	gingersnaps, crushed (if unavailable use graham crackers)	360 mL
1/4 cup	granulated sugar	60 mL
1/3 cup	butter, melted	80 mL

Preheat the oven to 350°F (175°C). Lightly butter a 9-inch (23-cm) springform pan.

Crush the gingersnaps in a food processor. Add the sugar and melted butter and process just until combined. Transfer to the prepared springform pan. Spread the crumb mixture over the bottom and 1 inch (2.5 cm) up the sides of the pan.

For the filling:

3	8-oz. (225-g) packages cream cheese, softened	3
1 cup	granulated sugar	240 mL
1/4 cup	brown sugar, packed	60 mL
1 3/4 cups	cooked pumpkin	420 mL
2	eggs	2
2 Tbsp.	whipping cream	30 mL
1 1/2 tsp.	ground cinnamon	7.5 mL
1/4 tsp.	ground nutmeg	1.2 mL
1/4 tsp.	ground ginger	1.2 mL

In a food processor or with an electric mixer, beat the cream cheese with the sugars until light and fluffy. Beat in the pumpkin, eggs, cream and spices. Pour over the crumb mixture and bake for 55 to 60 minutes, or until the cheesecake is firm and set. Remove from the oven.

For the topping:

2 cups	*sour cream*	*480 mL*
1/4 cup	*granulated sugar*	*60 mL*
1 tsp.	*vanilla extract*	*5 mL*

Mix the sour cream, sugar and vanilla extract together and pour over the warm cheesecake. Return the cake to the oven and bake for 5 minutes. Cool the cake on a wire rack.

Use a sharp knife dipped in hot water to free the edges of the cake before opening the pan. Chill for several hours or overnight before serving.

SWEET POTATO AND APPLE PIE WITH PECANS

For the crust:

1 cup	all purpose flour	240 mL
1 Tbsp.	granulated sugar	15 mL
1/2 tsp.	ground cinnamon	2.5 mL
1/2 tsp.	salt	2.5 mL
1/3 cup	vegetable shortening or lard	80 mL
1 Tbsp.	chilled butter	15 mL
2 Tbsp.	cold water	30 mL

Sift the flour, sugar, cinnamon and salt together in a large bowl. Cut in the shortening and butter with a pastry blender until the dough is pea-size. Sprinkle with the water and continue blending until all the ingredients hold together. Gather into a ball, flatten into a disc shape, wrap in plastic wrap and let rest in the refrigerator for 30 minutes.

Have ready a 9-inch (23-cm) pie plate. Roll out the pastry dough so it's 2 inches (5 cm) larger than the pie plate. Transfer the dough to the plate. Trim the edges and crimp the crust between thumb and forefinger to form a decorative edge. Place in the refrigerator while making the filling.

For the filling:

2 cups	sweet potato, cooked and mashed	480 mL
1/2 cup	brown sugar	120 mL
2 Tbsp.	rum	30 mL
1 Tbsp.	butter, melted	15 mL
1 tsp.	ground cinnamon	5 mL
1 tsp.	lemon rind	5 mL
1/2 tsp.	grated nutmeg	2.5 mL
1/4 tsp.	mace	1.2 mL
1/4 tsp.	salt	1.2 mL

2	eggs, separated	2
3/4 cup	milk	180 mL
1/2 cup	whipping cream	120 mL
1/2 cup	brown sugar	120 mL
1 Tbsp.	butter, melted	15 mL
1/2 tsp.	cinnamon	2.5 mL
1 1/2 cups	apples, peeled, cored and sliced	360 mL
1/2 cup	pecans, toasted (see page 183) and chopped	120 mL

Preheat the oven to 375°F (190°C).

With an electric mixer, combine the sweet potato, 1/2 cup (120 mL) of brown sugar, rum, 1 Tbsp. (15 mL) of melted butter, 1 tsp. (5 mL) of cinnamon, lemon rind, nutmeg, mace and salt in a medium bowl. Beat in the egg yolks one at a time until well combined. Add the milk and whipping cream and continue to beat until well mixed. Beat the egg whites until stiff. Fold them into the sweet potato mixture.

In a separate bowl, combine the remaining brown sugar, butter and cinnamon.

Arrange the apple slices on the bottom of the pie crust. Sprinkle with the brown sugar mixture. Scatter the pecans on top. Pour the sweet potato mixture over all. Bake for 20 minutes. Reduce the heat and bake for 45 minutes longer, or until the pie filling is set.

Cool completely before serving.

SPICY PUMPKIN PIE

For the crust:

1 cup	all purpose flour	240 mL
1 Tbsp.	granulated sugar	15 mL
1/2 tsp.	ground cinnamon	2.5 mL
1/2 tsp.	salt	2.5 mL
1/3 cup	vegetable shortening or lard, chilled	80 mL
1 Tbsp.	butter, chilled	15 mL
2 Tbsp.	cold water, more or less	30 mL

Serves 6 to 8

Hot or cold, any way you slice it, be sure to serve this pie with big dollops of cinnamon-scented whipped cream.

Sift the flour, granulated sugar, cinnamon and salt into a large bowl. Cut in the shortening or lard and butter with a pastry blender until the dough is pea-size. Sprinkle with the water and continue blending until all the ingredients hold together. Gather into a ball and flatten into a disc shape, wrap in plastic and let rest in the refrigerator for half an hour.

Have ready a 9-inch (23-cm) pie plate.

Roll out the pastry dough so it's 2 inches (5 cm) larger than the pie plate. Transfer the dough to the plate. Trim the edges and crimp the crust between thumb and forefinger to form a decorative edge. Place in the refrigerator while making the filling.

For the filling:

3	eggs	3
1/4 cup	brown sugar	60 mL
1/2 cup	granulated sugar	120 mL
2 cups	cooked pumpkin purée	480 mL
1 1/2 cups	sour cream	360 mL
1 1/2 tsp.	ground cinnamon	7.5 mL
1 tsp.	molasses	5 mL
1/2 tsp.	ground ginger	2.5 mL
1/4 tsp.	each allspice and ground cardamom	1.2 mL
1/8 tsp.	cloves	.5 mL
	pinch salt	

Preheat the oven to 425°F (220°C).

Beat the eggs and both sugars until light. Add the pumpkin, sour cream, cinnamon, molasses, ginger, allspice, cardamom, cloves and salt. Mix until well combined.

Pour the filling into the pie shell. Bake for 10 minutes. Reduce the heat to 350°F (175°C) and continue to bake 45 minutes longer, or until a tester inserted into the center comes out clean.

For the topping:

1 cup	*whipping cream*	*240 mL*
1 Tbsp.	*granulated sugar*	*15 mL*
1 tsp.	*vanilla*	*5 mL*
	dusting of cinnamon	

Whip the cream, sugar and vanilla until soft peaks form.

Serve the pie warm or cold, with mounds of whipped cream. Dust the cream with a little cinnamon.

Rhubarb Sour Cream Pie

For the crust:

1 cup	all purpose flour	240 mL
2 tsp.	granulated sugar	10 mL
1/2 tsp.	salt	2.5 mL
1/3 cup	shortening or lard, chilled	80 mL
1 Tbsp.	butter, chilled	15 mL
2 Tbsp.	water, more or less	30 mL

Serves 6

I didn't care much for rhubarb until I tried this recipe at my friend Diane's house.

Sift the flour, sugar, and salt together in a large bowl. Cut in the shortening or lard and butter with a pastry blender until the dough is pea-size. Sprinkle with the water and continue blending until all the ingredients hold together. Gather into a ball and flatten into a disc, wrap in plastic and let rest in the refrigerator for 30 minutes.

Have ready a 9-inch (23-cm) pie plate.

Roll out the pastry dough so it's 2 inches (5 cm) larger than the pie plate. Transfer the dough to the plate. Trim the edges and crimp the crust between thumb and forefinger to form a decorative edge.

For the filling:

4 cups	rhubarb, cut in 1/2-inch (1.2-cm) pieces	1 L
1 1/2 cups	granulated sugar	360 mL
1/3 cup	flour	80 mL
1 cup	sour cream	240 mL
1/2 cup	flour	120 mL
1/2 cup	brown sugar	120 mL
1/4 cup	butter	60 mL

Preheat the oven to 425°F (220°C). Arrange the rhubarb pieces over the bottom of the pie crust.

Combine the granulated sugar, 1/3 cup (80 mL) of flour and sour cream. Mix well and pour over the rhubarb.

Combine the 1/2 cup (120 mL) flour, the brown sugar and butter in a bowl. Using your fingertips, rub the mixture together until it forms coarse crumbs. Sprinkle the mixture over the sour cream and rhubarb mixture.

Bake for 15 minutes. Reduce the heat to 350°F (175°C) and continue to bake for 30 minutes longer, or until it's a light golden color.

Serve cold or at room temperature.

AUTUMN PUMPKIN AND MAPLE FLAN

Serves 6

This is a homey, comfort-food kind of dessert. It's even better served the next day.

4	eggs	4
1/2 cup	sugar	120 mL
1 1/2 cups	cooked pumpkin, mashed	360 mL
1/2 tsp.	ground cinnamon	2.5 mL
1/2 tsp.	ground ginger	2.5 mL
1/4 tsp.	grated nutmeg	1.2 mL
1/4 tsp.	salt	1.2 mL
1 cup	milk	240 mL
1 cup	half-and-half cream	240 mL
1 cup	whipping cream	240 mL
1 tsp.	sugar	5 mL
1/2 tsp.	vanilla	2.5 mL
1/2 cup	maple syrup	120 mL
	dusting of cinnamon	

Preheat the oven to 350°F (175°C). Lightly butter six 3/4-cup (180-mL) ramekins.

Beat the eggs with an electric mixer in a medium bowl. Gradually add the 1/2 cup (120 mL) sugar. Add the pumpkin, cinnamon, ginger, nutmeg and salt. Continue to mix until well combined.

Heat the milk and cream in a small saucepan until hot but not boiling. Slowly add it to the pumpkin mixture, stirring constantly.

Pour the pumpkin mixture into the prepared ramekins. Place the ramekins on a rack in a baking dish. Add hot water until it comes halfway up the sides of the ramekins. Cover with foil. Bake for 40 minutes, or until the flans are firm. Refrigerate until completely cooled.

Whip the cream, 1 tsp. (5 mL) sugar and vanilla until soft peaks form. Pour a little maple syrup on top of each flan. Place a dollop of whipping cream on top and dust with a little cinnamon.

PANTRY

*Having a few homemade items on hand can make
a big difference when you're cooking. Here are some
basic recipes for dressings, pestos, salsas and a few
home-canned condiments.*

*It may seem a little laborious when you're making them,
but nothing is more satisfying than reaching into the cupboard and retrieving a jar of homemade pickles or relish.*

*The end of summer is usually the best time for making
pickles and relishes. Vegetables are at the height of
freshness and they're plentiful and less expensive.*

*For the recipes requiring water bath canning,
please follow the instructions carefully.*

HOMEMADE MAYONNAISE

3	egg yolks	3
1 Tbsp.	Dijon mustard	15 mL
1 1/2 cups	vegetable oil	360 mL
2 tsp.	white wine vinegar or lemon juice	10 mL
	salt and freshly ground black pepper to taste	

Makes approximately 2 cups (480 mL)

In French cuisine, mayonnaise is known as a mother sauce. Adding ingredients creates a different sauce. Add relish and hard-boiled egg to make tartar sauce.

Pulse the egg yolks and mustard in a food processor until blended. While the motor is running, add the oil a few drops at a time, gradually increasing to a slow steady stream. When the mixture has thickened, add the vinegar or lemon juice, salt and pepper. Continue blending until it's well mixed and the color has gone from yellow to pale white.

If the dressing is too thick, add a little chicken stock or water.

Mayonnaise can be kept in an airtight container in the refrigerator for up to 1 week.

CRÈME FRAÎCHE

Makes 2 cups (480 mL)

| 1 cup | whipping cream | 240 mL |
| 1 cup | sour cream | 240 mL |

Serve over fruit, as a simple sauce for hot or cold vegetables or added to soup just before serving to give it body and extra flavor.

In a bowl whisk together both creams until well combined. Cover the mixture with plastic wrap and let it sit at room temperature for 24 hours. Refrigerate at least 2 hours before serving.

It will keep in the refrigerator for up to 2 weeks. When you're planning to have guests, make it a few days ahead.

BASIC VINAIGRETTE

1/4 cup	vinegar or lemon juice	60 mL
	salt and freshly ground black pepper to taste	
1	shallot, finely minced	1
3/4 cup	extra virgin olive oil	180 mL
1 tsp.	Dijon mustard (optional)	5 mL

*Makes 1 cup
(240 mL)*

*This is the basic recipe
for any vinaigrette.
Change the flavor by
changing the vinegar.*

Dissolve the salt and pepper by whisking into the vinegar or lemon juice. Add the shallot and olive oil and whisk until combined. Whisk in the mustard, if desired.

SPICED GRILLING OIL FOR VEGETABLES

3/4 cup	vegetable oil	180 mL
2 Tbsp.	balsamic vinegar	30 mL
1	clove garlic, finely chopped	1
1/4 tsp.	salt	1.2 mL
1/4 tsp.	freshly ground black pepper	1.2 mL
1/4 tsp.	crushed chili peppers (optional)	1.2 mL

*Makes 3/4 cup
(180 mL)*

*When grilling vegetables
on the barbecue, a baste
of flavored vegetable
oil helps prevent the
vegetables from sticking
as well as adding
extra flavor.*

Combine all the ingredients in a bowl and brush onto the vegetables before and during grilling. Grill the vegetables over medium heat. Try using fresh herbs, such as thyme or rosemary, tied together to make a basting brush. Brush the vegetables with oil again just before serving. The optional crushed chili peppers add a little fire to the baste.

SUN-DRIED TOMATO PESTO

1 cup	sun-dried tomatoes	240 mL
6	cloves garlic	6
2 cups	fresh basil leaves, packed	480 mL
1/2 cup	fresh grated Parmesan cheese	120 mL
1/4 cup	pine nuts	60 mL
1 cup	extra virgin olive oil	240 mL

*Makes about
2 cups (480 mL)*

*Sun-dried tomatoes give
this pesto a more subtle
flavor. Use it anywhere
you would use Basil Pesto
(below). Freeze leftover
portions to use later.*

Bring a small saucepan of water to a boil, add the sun-dried tomatoes and simmer for about 3 minutes, until the tomatoes are just tender. Drain.

In a food processor finely chop the tomatoes, garlic, basil, Parmesan cheese and pine nuts.

With the motor running, gradually add the olive oil in a thin steady stream.

BASIL PESTO

2 cups	fresh basil leaves, packed	480 mL
3	cloves garlic	3
1/2 cup	freshly grated Parmesan cheese	120 mL
1/4 cup	pine nuts	60 mL
1/2 tsp.	lemon zest	2.5 mL
3/4 cup	extra virgin olive oil	180 mL

*Makes about 1 cup
(240 mL)*

*Use pesto with pasta, to
zip up a soup, to stuff
chicken breasts and
pork roasts.*

In a food processor finely chop the basil, garlic, Parmesan cheese, pine nuts and lemon zest.

With the motor running, gradually add the oil in a thin steady stream.

Southwest Tomato Salsa

4	medium tomatoes	4
1	medium onion, finely chopped	1
1/2	sweet red pepper, chopped	1/2
1	jalapeño pepper, seeded and chopped	1
3	cloves garlic, minced	3
1/2 cup	tomato sauce	120 mL
1/2 cup	fresh or frozen corn	120 mL
1/3 cup	lemon juice	80 mL
1/4 cup	chopped fresh parsley	60 mL
1/4 cup	chopped fresh cilantro	60 mL
2 tsp.	sugar	10 mL
1 tsp.	cumin seed	5 mL
1/2 tsp.	salt	2.5 mL
1/2 tsp.	mustard seed	2.5 mL

*Makes 3 cups
(720 mL)*

This salsa is best in summer when fresh tomatoes are full of flavor, but you can substitute canned tomatoes.

Bring a large pot of water to a boil. Make an X in the bottom of each tomato. Drop them in boiling water for 1 minute, then plunge them into ice cold water. The skins will peel off easily. Chop the tomatoes.

Combine all the ingredients in a large saucepan over medium heat. Bring to a boil, reduce the heat and simmer for 20 minutes or until the sauce is thick. Refrigerate for up to 2 weeks.

BLACK BEAN WITH PINEAPPLE SALSA

1 cup	dried black beans	240 mL
1 cup	chopped fresh or canned pineapple	240 mL
1 cup	diced red onion	240 mL
1 Tbsp.	finely chopped jalapeño pepper	15 mL
1/2 cup	chopped fresh coriander	120 mL
1/4 cup	lime juice	60 mL
1 tsp.	grated lime zest	5 mL
1/4 cup	extra virgin olive oil	60 mL
	salt and freshly ground black pepper to taste	

Makes approximately 4 cups (1 L)

Salsas hail from Latin America, where they have long been used to excite the palate. Somewhere between a salad and a sauce in texture, this salsa is a great accompaniment to grilled fish or seafood.

Place the black beans in a medium saucepan, cover with approximately 2 1/2 cups (600 mL) of water and bring to a boil. Reduce the heat to low and simmer until the beans are tender, about 45 minutes. Drain well. Cool slightly.

Combine the cooled beans and all the other ingredients in a bowl. Refrigerate at least 1 hour before serving.

VIDALIA ONION AND THREE-PEPPER RELISH

2	medium Vidalia onions, finely chopped	2
1	large cooking onion, finely chopped	1
1	jalapeño pepper, seeds removed and finely chopped	1
1	sweet green pepper, finely chopped	1
1	sweet red pepper, finely chopped	1
3	large cucumbers, peeled, seeded and finely chopped	3
2 Tbsp.	pickling salt	30 mL
1 1/4 cups	cold water	300 mL
1 1/2 cups	sugar	360 mL
1/4 cup	all purpose flour	60 mL
2 Tbsp.	dry mustard	30 mL
3/4 tsp.	turmeric	4 mL
3/4 tsp.	mustard seed	4 mL
3/4 tsp.	celery seed	4 mL
1 1/2 cups	white vinegar	360 mL
1/2 cup	water	120 mL

Makes 6 1-cup (240-mL) jars

Don't let the peppers scare you; this is a sweet golden relish that's great with burgers, sausages and cold meats. It can turn a humdrum hot dog into a swanky frank.

Combine all the vegetables in a large bowl. Sprinkle with the salt and add the 1 1/4 cups (300 mL) of cold water. Let stand for 1 hour. Drain well.

In a large heavy saucepan, mix together the sugar, flour, dry mustard, turmeric, mustard seed and celery seed. Gradually stir in the vinegar and the 1/2 cup (120 mL) of water.

Add the drained vegetables and bring to a boil. Simmer for about 45 minutes, stirring often, until thickened. Pack into hot, sterilized, 1-cup (240-mL) jars, leaving 1/2 inch (1.2 cm) head room. (See Instructions for Water Bath Canning, page 221.) Wipe the rims and apply the lids. Process in a water bath for 15 minutes.

PICKLED ASPARAGUS SPEARS

2 1/2 lbs.	asparagus spears	1.1 kg
4	small jalapeño or chili peppers	4
4	cloves garlic	4
1 Tbsp.	mustard seed	15 mL
8	whole black peppercorns	8
2 cups	white wine vinegar	480 mL
1 cup	dry white wine	240 mL
3/4 cup	water	180 mL
3 Tbsp.	sugar	45 mL
1 tsp.	pickling salt	5 mL

*Makes 4 pint
(500-mL) jars*

*When you've had your
fill of fresh asparagus,
here's an interesting way
to preserve the harvest for
later in the year. Select
unblemished, firm,
straight spears of about
the same thickness with
closed, dry tips. Save the
unused portion of the
stalks for making Cream
of Asparagus Soup
(page 38) or Asparagus
Flans (page 100).*

Wash the asparagus thoroughly to remove any sand. Cut the
spears to fit in your jars. For pint (500-mL) jars, they should be
approximately 4 1/4 inches (11 cm), leaving a 1/2-inch (1.2-cm)
head space in the jar. Soak the asparagus in ice water for 1 hour.

Sterilize 4 pint (500-mL) jars in boiling water. (See Instructions for
Water Bath Canning, page 221.)

Into each sterilized jar place 1 jalapeño or chili pepper, 1 clove
garlic, 1/4 of the mustard seed and 2 black peppercorns. Pack the
asparagus in the jars with the tips down.

Combine the vinegar, wine, water, sugar and salt in a saucepan
and bring to a boil. Boil for 5 minutes. Pour the hot liquid over the
asparagus, leaving 1/2 inch (1.2 cm) head room. Wipe the rims
and apply the lids. Place the jars in the canner and process for
15 minutes.

Allow the asparagus pickles to age for 2 weeks before using them.

INSTRUCTIONS FOR WATER BATH CANNING

1. Fill a water bath canner with water and bring it to a boil.

2. Sterilize clean jars by placing them in boiling water for 10 minutes. Do not let the jars touch each other.

3. Place the canning lids in a small pan of boiling water. Boil them for 5 minutes to soften the sealing compound.

4. Place the food into the sterilized jars. Remove air bubbles by sliding a sterilized rubber spatula between the jar and food. Readjust the head space. Wipe the rim with a clean cloth. Center the snap lid and apply the screw band.

5. Place the jars in the canner, making sure the water covers the tops of the jars by 2 inches (5 cm). Bring the water to a boil and boil for the length of time asked for in the recipe. Cool the jars for 24 hours. Check to see that they are sealed: sealed lids curve downward. Wipe the jars and store them in a cool dark place.

NAN'S CHILI SAUCE

6 quarts	*very ripe field tomatoes*	6 L
3	*sweet red peppers, finely chopped*	3
3	*sweet green peppers, finely chopped*	3
6	*large apples, cored and diced*	6
4	*large onions, finely chopped*	4
4 cups	*white vinegar*	1 L
1	*small bunch celery, finely chopped*	1
2 lbs.	*brown sugar*	900 g
2 Tbsp.	*salt*	30 mL
1 tsp.	*cayenne pepper*	5 mL
1 tsp.	*ground allspice*	5 mL
1 tsp.	*ground cloves*	5 mL

*Makes 14 pint
(500-mL) jars*

*This is my grandmother's
recipe, which has been in
our family for as long as
anyone can remember.
My grandmother is no
longer with us, but every
fall she's there in my
kitchen as I carry on the
fall ritual. We use chili
sauce in place of ketchup
— on egg dishes, on
burgers, and alongside
meat pies.*

Bring a large pot of water to a boil. Make an X in the bottom of each tomato. Drop them in boiling water for 1 minute, then plunge them into ice cold water. The skins will peel off easily. (If the skins do not peel off easily, boil them a little longer.) Place the tomatoes in a large colander to drain while chopping the rest of the vegetables. If you have a food processor, use it to chop all the vegetables except the tomatoes; it saves a lot of time. Just be sure not to purée them.

Transfer the tomatoes to a preserving kettle and crush them with a potato masher.

Place the preserving kettle over medium heat, add the rest of the ingredients and bring the sauce to a boil. Reduce the heat to low and simmer for 2 1/2 to 3 hours, or until the mixture thickens. Be sure to stir the sauce frequently to prevent scorching. Remove any scum that comes to the top of the sauce.

Sterilize 14 pint (500-mL) jars in boiling water. (See Instructions for Water Bath Canning, page 221.)

Ladle the sauce into the hot sterilized jars, leaving 1/2 inch (1.2 cm) head room. Wipe the rims and apply the lids. Process in a water bath for 20 minutes.

Sweet Red and Jalapeño Pepper Jelly

5 cups	*sugar*	1.2 L
2 cups	*sweet red peppers, finely chopped*	480 mL
1	*jalapeño pepper, seeds removed and finely chopped*	1
1 1/2 cups	*white vinegar*	360 mL
1	*6-oz. (170-mL) bottle liquid pectin*	1

Combine the sugar, red peppers, jalapeño pepper and vinegar in a large saucepan. Bring to a boil over medium-high heat and continue to boil for 15 minutes, removing any scum that comes to the surface.

Add the pectin and continue to boil for 1 minute. Remove from the heat and stir 1 minute more.

Pour the jelly into hot sterilized jars, leaving 1/2 inch (1.2 cm) head room. (See Instructions for Water Bath Canning, page 221.) Wipe the rims and apply the lids. Process in a water bath for 10 minutes.

Note: You can chop the sweet and jalapeño peppers in a food processor, but do not purée them. Puréeing will make the jelly cloudy.

Makes 6 1-cup (240-mL) jars

This jelly is very easy to make and goes well with everything, especially cheese. For a simple hors d'oeuvre to serve with crackers, prick a wheel of Brie in a few places with a fork, spoon a little pepper jelly over it and just warm the Brie in the microwave.

INDEX

ABOUT THE AUTHOR

Fed up with the conventional 9 to 5, Darlene King followed her dream and enrolled in the Cuisine program at Le Cordon Bleu Cooking School of Paris in Ottawa, Ontario. She has since worked as a chef, food writer and cooking instructor, but is best known for her efforts as food editor at *Harrowsmith Country Life* magazine. This cookbook, her first, was born of one of her many articles. She lives in the Northumberland Hills of Ontario with her husband, Tom, and a golden retriever named Scout.